Contents

Wild Flowers

OF THE VICTORIA FALLS AREA

CREDITS

Photographs: Helen Pickering and Evelyn Roe

Identifications: Mike Bingham and Mark Hyde

Text: Helen Pickering and Evelyn Roe

Map: Bernd Eversmann

Book Design: Michael Sturley

Website: www.kalahariplants.webs.com

First published by Helen Pickering, London, 2009.

ISBN: 978-0-9563634-0-4

Printed by Aquila Printers Ltd, Lusaka, Zambia.

Acknowledgements

The authors would like to thank Oliver Sikatumba, field assistant, whose sharp eyes and helpful suggestions added greatly to the scope of this book; in Livingstone we are indebted to Michael Nang'alelwa from The National Heritage Conservation Commission, Clare Mateke, Freddie Siangulube and Henry Muloongo from The Livingstone Museum; Bob Stjernstedt, Cliff Sitwell, Ali Shenton, Mike Musgrave and Stuart Norman, members of the Livingstone branch of the Wildlife and Environmental Conservation Society of Zambia, Brett Saunders and Matusa Kembule all of whom assisted with field data collection and identifications.

In Harare we would like to thank Meg Coates-Palgrave for her advice on shrub identifications and Christopher Chapano, Keeper of the Herbarium.

In London we would like to thank Jane Browning, for her help with orchid identification, as well as Jonathan Timberlake, Iain Darbyshire and Shahina Ghazanfar at RBG, Kew for their support and advice.

To dear Flea

Let's go and look at these flowers together sometime soon.

with much love,

Evelyn

February 2010

The Victoria Falls

Mosi-oa-Tunya or ***'The Smoke that Thunders'*** can be seen from over 30 kilometres away, billowing into the air, creating a focal point in an otherwise rather uniform landscape.

The Zambezi River runs for approximately 2740 kilometres with the Victoria Falls situated almost half way between its source, in a marshy bog in north-west Zambia, and its mouth at the Indian Ocean. Immediately above the Falls (altitude 915 metres) the river is nearly 1800 metres wide. It then plunges 110 metres into a gorge which, at its narrowest point, is only 60 meters across. The river then zigzags through a series of deep gorges, which mark the positions of previous Falls, over a series of rapids, widening out into the Kariba dam and finally through Mozambique to the sea.

History

Historically the Zambezi valley, occupied since the early Stone Age, has had a low population density, partly due to the presence of malaria and sleeping sickness, both of which are less common on the higher Central African plateau.

The first European to describe the Falls was David Livingstone in the late 1850s. They soon became a magnet that attracted explorers, naturalists and adventurers of all kinds.

In 1904 Cecil Rhodes commissioned the bridge across the second gorge as part of his vision to connect The Cape to Cairo. With the opening of the bridge, the town of Livingstone, in Zambia, became the regional administrative centre.

Climate

Climatically there are four distinct seasons:

The hot, dry season from September to October; when the mean maximum temperature is 33.9°C, rising, at times, above 40°C.

The rainy season from November to March and early April; annual rainfall varies from year to year, but is in the range 700 to 800 mm.

The early dry season from April to May; which is dry and fairly warm.

The winter season from June to August; is cool and dry. In June/July, the mean minimum temperature is 6.4°C, night-time temperatures may drop close to zero.

There is considerable seasonal variation in the volume of water flowing in the Zambezi and, hence, the amount of spray at the Falls. From January to June peak flows are 550,000 cubic metres per minute, producing a permanent cloud of spray.

Between June and December, the spray is considerably reduced. The rainy season begins in November, but this is the time when the Falls are at their lowest level, with a mean flow of only 20,000 cubic metres per minute.

Geology

The underlying rocks are the product of volcanic action that took place about 150 - 200 million years ago, resulting in lava flows forming a basalt plate. The Falls themselves are formed by the waters of the Zambezi falling over a vertical-sided chasm and eroding the rock of the river-bed.

The terrain consists mainly of shallow, basaltic soils overlain, close to the river, by alluvial sand and gravel, and by red Kalahari sands (wind-blown) further away. The many islands upstream are either basalt outcrops with thin soil, or alluvial deposits covered withreeds, grasses and mature trees.

Vegetation

There are two main vegetation zones in the vicinity of Victoria Falls: the riverine forest close to the river and semi-deciduous open woodland further away.

In the immediate area of the Falls there is a small area of closed forest, popularly known as the 'rainforest', within the spray zone. This has the typical three layers of rainforests: a canopy of large trees, an intermediate shrub layer and an herbaceous layer consisting of climbers, weedy plants and ferns, which is closely intertwined with the shrub layer.

In common with other African rivers, the Zambezi has a fringe of forest along its entire length, which, in places, may be only one tree deep. In addition, patches of riverine forest can be found in the gorges, especially where tributary streams join the Zambezi, and on the islands.

Colophospermum mopane

Away from the river the vegetation varies with the soil: on the alluvial sands and gravel it is known as Mopane; characterised by *Colophospermun mopane*, with its butterfly-shaped leaves, and the easily recognisable Baobab *(Adansonia digitata)*.

Adansonia digitata

Schinozophyton rautanenii

Baikiaea plurijuga

Areas of deep, well-drained Kalahari sands have mainly open deciduous woodland characterised by *Schinziophyton rautanenii*, (Mongongo) a highly conspicuous tree in the dry season, and *Baikiaea plurijuga* (Zambian Teak), a hard wood, used for railway sleepers, flooring and furniture, with a reddish bark.

Organisation of this guide

This guide is designed as a quick reference for both amateur and professional plant lovers. Owing to the large number of books available on the trees of southern Africa we have focused on herbaceous plants and shrubs up to about 2 meters high. 220 species are illustrated and described briefly. We have included only features that can be seen by the naked eye. The book is organised alphabetically by family (as found in Flowering Plant Families of the World, Kew), genus and species. Identifications do not go beyond species level.

Nine locations were visited in all seasons, attempting to represent the range of vegetation types found around the Victoria Falls, the Gorges below the Falls, and up to 40km upstream on the Zambian side of the river. These are shown as areas 1 to 9 on the map (opposite). Below is a description of each of these sites.

Area 1 *The fenced area of the Falls.* This contains firstly, the spray zone which for many months of the year immerses visitors in such dense spray that they can hardly see either the waterfalls or the dramatic Batoka Gorge. In addition there are paths down into the Boiling Pot, and up and down stream that show mainly riverine vegetation with some weeds in disturbed areas further away from the river.

Area 2 *Gorge swing*. Here there are various activities such as a wire swing over the gorge and abseiling. A good path with thick, mainly riverine, vegetation leads down to the rocky edge of the river.

Area 3 *Songwe gorge*, also known as gorge 10. A very steep climb down over exposed basalt rocks where vegetation is limited to plants which require little water.

Area 4 *Taita Falcon gorge.* Here a small tributary joins the Zambezi allowing a wider range of trees, shrubs and herbaceous plants to grow all year round.

Area 5 *Mukuni ridge.* An area of both alluvial sand/gravel and Kalahari sand which has a wide variety of herbaceous flora.

Area 6 *Dambwa quarry.* An open area of Kalahari sand with many herbaceous and weedy plants. Beyond the quarry there is an area of black clay soil with Acacia woodland, **6a.**

Area 7 *The confluence of the Sinde river with the Zambezi.* Close to the Zambezi there is a rich, mainly riverine, flora on alluvial sand; further away Mopane scrub dominates.

Area 8 *Bovu Island and nearby villages.* The dirt road to Bovu passes through Kalahari sand then drops down to Mopane scrub. Around the villages there is a variety of weedy species. The island provides a very good example of riverine vegetation and water plants.

Area 9 *The environs of Livingstone.* Includes roadsides, edges of the railway line, the sewerage ponds and areas around hotels and lodges. In these locations we find many pan-tropical weeds.

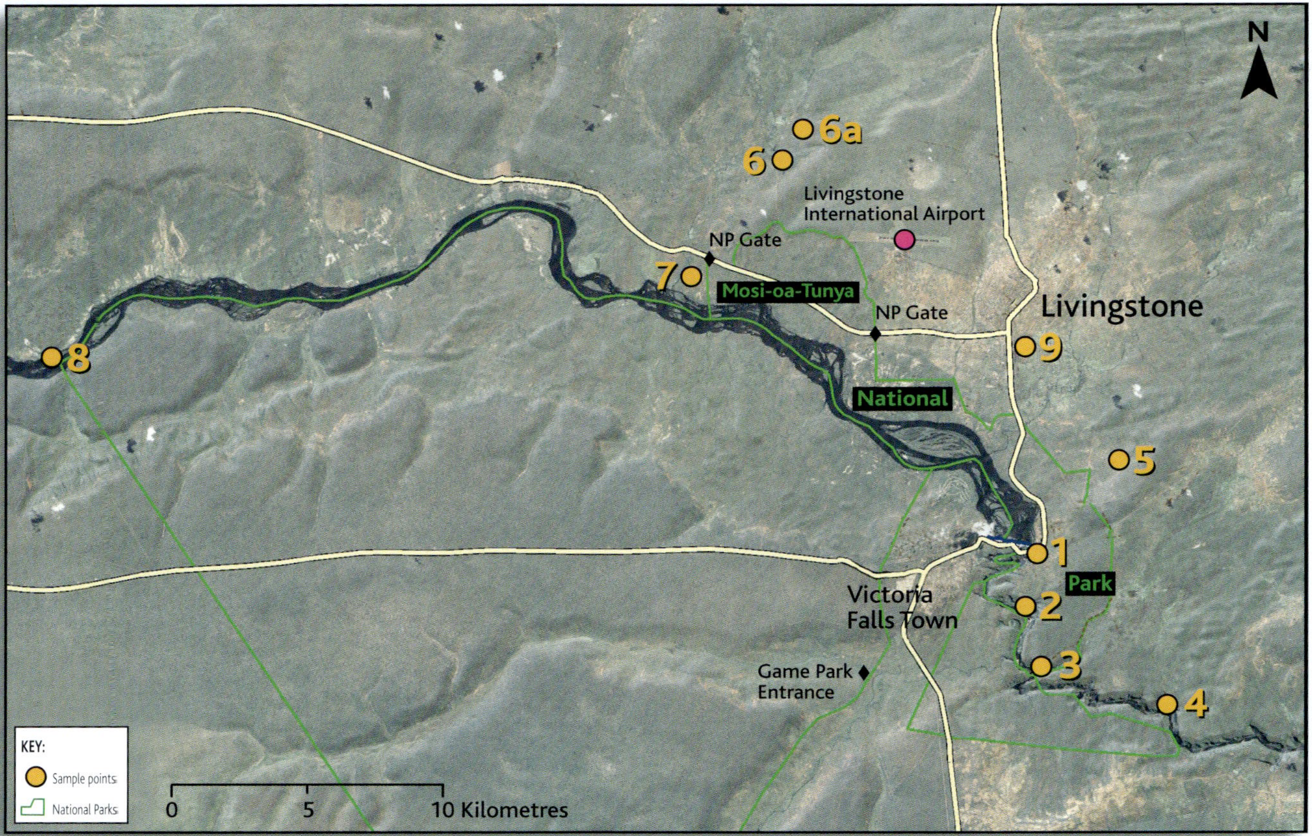

© Provincial centre for Geographical Information Services (PCGIS)

Acanthaceae (ACANTHUS FAMILY)

NAME *Asystasia gangetica*

A fast-growing, upright perennial; up to 1 m. *Leaves* Opposite, dark green, simple, oval with entire margins. *Flowers* White with purple markings on the lower lip. *Fruit* Capsule with brown seeds.

Flowering	Rains.
Habitat	Open grassland; areas 1, 7 and 8.
Distribution	Throughout the Old World tropics, with a number of sub-species.
Note	Pollinated by honey bees. Attracts butterflies.

NAME *Asystasia mysorensis*

Erect annual, up to 45 cm. *Leaves* Opposite, elliptic, slightly hairy. *Flowers* White with green markings emerging from long, hairy bracts.

Flowering	Rains.
Habitat	Weed in disturbed areas, generally found in damp locations; area 1.
Distribution	Southern and East Africa.
Note	Used as a vegetable.

NAME *Barleria kirkii*

Shrub; up to 1.5 m. *Leaves* Opposite, rounded, hairy with entire margins. *Flowers* Mauve, with five free petals, trumpet-shaped with protruding stamens.

Flowering Late rains to early dry season.

Habitat Dry, rocky or sandy places; areas 2 and 9.

Distribution Central Africa.

NAME *Barleria matopensis*

An evergreen shrub with hairy stems: up to 75 cm. *Leaves* Opposite, pale green, oval with entire margins. *Flowers* Pale mauve with a darker throat, trumpet-shaped with protruding stamens. *Fruit* An exploding capsule.

Flowering Rains.

Habita Damp, shady places; area 1.

Distribution Old World tropics.

Note Pollinated by several species of insects, attracts butterflies and birds.

Acanthaceae (ACANTHUS FAMILY)

NAME *Blepharis maderaspatensis*

A spiny, procumbent herb with hairy stems; up to 20 cm. *Leaves* Whorls of 3-4 leaves of unequal size, oval with entire margins. *Flowers* Mauve, or white with mauve stripes, bracts spiny. *Fruit* Mature fruits have hooked papery bracts that stick to clothing.

Flowering	Rains.
Habitat	Weed of wasteland and along edges of roads; area 9.
Distribution	Tropical Africa including Madagascar, extending to Pakistan and India.
Note	Used medicinally in West Africa for gout and oedema.

NAME *Crabbea velutina*

An erect perennial; up to 25 cm. *Leaves* Erect, arising from the base, oblong with smooth margins. *Flowers* White with a yellow centre.

Flowering	Rains.
Habitat	Common in the rainforest. Also found in dry grassland; area 1.
Distribution	East, Central and southern Africa.

NAME *Duosperma crenatum*

Erect, branched herb with hairy quadrangular stems; up to 1.2 m. *Leaves* Opposite, with entire margins at base, slightly serrated at the top. *Flowers* White with purple markings, in groups at intervals along the stem, surrounded by pale green bracts. *Fruit* A shiny brown flattened capsule.

Flowering	Rains.
Habitat	Damp, shady places, may form thickets at forest margins; area 1.
Distribution	East and Central Africa.
Note	Food plant for many butterflies. Young leaves cooked as a vegetable. Old, yellowing leaves believed to be toxic to humans.

NAME *Hygrophila abyssinica*

Mainly prostrate herb; up to 20 cm. *Leaves* Opposite, oval with smooth margins. *Flowers* Usually blue, sometimes white, with 5 free petals.

Flowering	Dry season.
Habitat	Along rivers and damp places; area 1.
Distribution	Throughout Africa.

NAME *Justicia heterocarpa*

An erect, hairy annual with a ribbed stem; up to 60 cm. *Leaves* Opposite, oval with soft hairs. *Flowers* White or pink with darker markings, surrounded by bracts. *Fruit* A dry capsule with four seeds.

Flowering Rains.

Habitat Locally common in dry grassland, often under shade; area 1.

Distribution Central, East and southern Africa.

Note Leaves are edible when cooked.

NAME *Justicia matammensis*

An erect annual with hairy stems; up to 40 cm. *Leaves* Opposite, hairy, oval with entire margins. *Flowers* White with pink markings, borne on spikes between bracts. *Fruit* A dry capsule.

Flowering Rains.

Habitat Sandy or disturbed areas of dry scrubland; area 7.

Distribution East and Central Africa.

Acanthaceae (ACANTHUS FAMILY)

NAME *Monechma debile*

Erect herb with angular ribbed stems; up to 2 m. *Leaves* Opposite, narrow with entire margins and conspicuous veins. *Flowers* Mauve or white, in hairy rounded bracts along the stem. *Fruit* A downy elongated capsule.

Flowering Late rains.

Habitat Wide variety of habitats, especially disturbed ground; area 9.

Distribution Throughout Africa.

NAME *Nelsonia canescens*

Prostrate branching herb; up to 50 cm. *Leaves* Opposite, oval, hairy with entire margins. *Flowers* Blue with five petals.

Flowering Dry season.

Habitat Open woodland and damp places; areas 1 and 9.

Distribution Widespread in the tropics.

Note Leaves have analgesic and anti-inflammatory properties. Plant used as a salt substitute in some countries.

NAME *Phaulopsis imbricata*

A scruffy, semi-erect herb; up to 1 m. *Leaves* Opposite, asymmetrical at the base with smooth margins. *Flowers* White, in large, green bracts on terminal spikes, sticky with an unpleasant odour.

Flowering Dry season.

Habitat Widespread in forests and woodland margins, very common; area 1.

Distribution Tropical and temperate regions of the Old World.

NAME *Ruellia cordata*

A shrubby perennial forming dense groups; up to 1.2 m. *Leaves* Opposite, oval with entire margins. *Flowers* Mauve trumpet-shaped opening into five lobes. *Fruit* Seeds slimy when wet.

Flowering Rains.

Habitat Damp shade; areas 1 and 8..

Distribution Central and southern Africa.

Note Named after Jean de la Ruelle, 16th century botanist, author of de Natura Plantarum, and a physician to Francis I of France.

Acanthaceae (ACANTHUS FAMILY)

NAME *Thunbergia alata*

A perennial climber; up to 1.2 m. *Leaves* Opposite, heart-shaped with entire margins. *Flowers* Yellow/orange with a dark purple centre. *Fruit* Round capsule with beak.

Flowering	Most of the year.
Habitat	Wide range of habitats, often near habitation; area 9.
Distribution	Native to East Africa, now a pan-tropical weed, invasive in Hawaii and Australia.
Note	Ornamental. In East Africa used medicinally, mainly for skin problems.

Aizoaceae (CARPET WEED AND NEW ZEALAND SPINACH FAMILY)

NAME *Sesuvium nyasicum*

A semi-prostrate, succulent annual; up to 20 cm. *Leaves* Alternate or sub-opposite, oblong, somewhat fleshy with entire margins. *Flowers* Solitary, pink or mauve with green tips. *Fruit* A conical capsule with black seeds.

Flowering	Rains.
Habitat	Open woodland; area 9.
Distribution	Central and southern Africa.
Note	Similar to *S. hydaspicum*.

NAME *Achyranthes aspera*

An erect annual; up to 1 m. *Leaves* Opposite, oval with entire margins. *Flowers* Cream/pink on an elongated hairy spike.

Flowering Early dry season.

Habitat Common in disturbed places; areas 1 and 8.

Distribution Pan-tropical weed.

Note Seeds are edible when cooked, have high protein content. Numerous uses in Ayurvedic medicine.

NAME *Aerva leucura*

Woody perennial, much branched, stems and leaves covered in small white hairs; up to 1 m. *Leaves* Alternate, elliptic with entire margins, turning yellow or red in autumn. *Flowers* Spikes with dense clusters of white, woolly bracts with tiny green flowers. *Fruit* Capsule.

Flowering Most of the year.

Habitat A wide variety of soils, common on disturbed ground; area 9.

Distribution Old World tropics.

Note Young leaves cooked as spinach. Flowering spikes used to stuff pillows.

Amaranthaceae (AMARANTHUS, CELOSIA & COCKSCOMB FAMILY)

NAME *Amaranthus hybridus*

Erect annual with red stems; up to 2.5 m. *Leaves* Alternate, dull green, oval with entire margins, on long stalks. *Flowers* Red, in clusters on long spikes, borne in the leaf axils.

Flowering Rains.

Habitat Disturbed ground, requires full sun; areas 1 and 9.

Distribution Native to tropical America, now pan-tropical.

Note If grown inorganically can concentrate nitrates from chemical fertilisers in the leaves which may poison cattle. There are several species of *Amaranthus* in the region.

NAME *Amaranthus spinosus*

Coarse erect herb with spines at the leaf nodes; up to 80 cm. *Leaves* Alternate, dull green, oval with entire margins. *Flowers* Tiny, green/red in dense clusters on long spikes in the leaf axils.

Flowering Rains.

Habitat Light soils in sunny locations; area 9.

Distribution Native to the Americas, now cosmopolitan.

Note From Greek 'amarantos' meaning not fading/everlasting. Valued as a vegetable in Africa. In Cambodia used as a grey dye for cloth.

Amaranthaceae (AMARANTHUS, CELOSIA & COCKSCOMB FAMILY)

NAME *Celosia trigyna*

Annual herb with green or pinkish stems; up to 1.8 m. *Leaves* Alternate, with entire margins. *Flowers* Clusters of pink/white flowers along the stems. *Fruit* Ovoid capsule with shiny black seeds.

Flowering Most of the year.

Habitat Widespread in disturbed areas, along forest tracks and the edges of plantations; areas 5 and 9.

Distribution Tropical Africa, including Madagascar, recorded in Arabia and as a weed in Florida.

NAME *Cyathula orthacantha*

Erect, branching hairy annual; up to 1 m. *Leaves* Opposite, variable shape with entire margins. *Flowers* Whitish/green spike at the end of branches with very small yellow or reddish flowers embedded in greyish bracts. *Fruit* Capsule with brown seeds.

Flowering Rains.

Habitat Disturbed ground and open woodland; area 9.

Distribution Common throughout tropical Africa.

NAME *Gomphrena celosioides*

A prostrate or ascending, hairy annual; up to 40 cm. *Leaves* Opposite, oblong with entire margins, white hairs along the margins and on underside. *Flowers* Small, yellow surrounded by conspicuous white bracts, in cylindrical spikes, lower flowers falling. *Fruit* A thin walled non-fleshy capsule, with brown shiny seeds.

Flowering	Most of the year.
Habitat	Widely distributed in disturbed and open ground; areas 1 and 9.
Distribution	Native to South America, now common in all tropical and sub-tropical regions.
Note	Poisonous to horses if eaten in large quantities.

NAME *Pupalia lappacea*

An erect annual or perennial covered with small hairs; up to 90 cm. *Leaves* Opposite, variable in size and shape with entire margins and a pointed tip. *Flowers* Woolly clusters on erect spikes with protruding white or pink/red hooked bracts. *Fruit* Ovoid capsule, shed with the flowers as a very sticky burr.

Flowering	Most of the year.
Habitat	Along roadsides, disturbed and rocky ground; area 1.
Distribution	Widespread in the Old World tropics and sub-tropics.

Amaryllidaceae (DAFFODIL FAMILY)

NAME *Crinum crassicaule*

Bulbous perennial; up to 50 cm. *Leaves* Basal rosette, linear with wavy margins. *Flowers* Clusters of white trumpet-shaped flowers with pink stripes, soon fall and hang downwards, fragrant.

Flowering	Early rains.
Habitat	Open areas or light woodland; areas 5 and 9.
Distribution	Sub-Saharan Africa.
Note	Bulbs have anti-viral properties.

NAME *Pancratium tenuifolium*

Bulbous perennial; up to 40 cm. *Leaves* Linear, appear just before the flowers. *Flowers* White, open at night and wilt by midday, fragrant.

Flowering	Early rains.
Habitat	Open areas or light woodland; areas 7 and 9.
Distribution	Sub-Saharan Africa.
Note	Bulbs may be hallucinogenic.

Amaryllidaceae (DAFFODIL FAMILY)

NAME *Scadoxus multiflorus*

A common perennial; up to 40 cm. *Leaves* Basal rosette, overlapping, dark green, arrow-shaped. *Flowers* Red, large spherical flower-heads of up to 200 star-shaped flowers.

Flowering Early rains.

Habitat Open areas or light woodland; areas 2 and 9.

Distribution Common throughout tropical Africa.

Annonaceae (SWEETSOP AND SOURSOP FAMILY)

NAME *Friesodielsia obovata*

A scrambling shrub with smooth bark; up to 7 m. *Leaves* Alternate, with conspicuous pale veins and smooth margins. *Flowers* Creamy yellow, slightly fleshy, solitary, in the leaf axils or terminally. *Fruit* Cylindrical, red, hairy. *Seeds* yellow.

Flowering Early rains.

Habitat Damp, shady areas, on termite mounds and rocky granite outcrops; areas 1, 8 and 9.

Distribution Central and southern Africa.

Note Fruit is edible.

Anthericaceae (SPIDER PLANT FAMILY)

NAME *Chlorophytum galpinii*

Robust perennial herb; up to 1.5 m. *Leaves* In a basal tuft, linear with entire margins, upright. *Flowers* White, large with long, yellow anthers. *Fruit* A capsule.

Flowering Rains.

Habitat Forest margins and open grassland; area 2.

Distribution Southern Africa.

NAME *Chlorophytum subpetiolatum*

Perennial herb; up to 40 cm. *Leaves* Basal, linear with smooth margins. *Flowers* White, clustered along a spike. *Fruit* Capsule with shallow lobes.

Flowering Early rains.

Habitat Open woodland, common; areas 1 and 9.

Distribution Sub-Saharan Africa.

NAME *Carissa edulis*

Scrambling shrub with opposite spines and milky sap; up to 5 m. *Leaves* Opposite, oval with entire margins, young leaves may be hairy. *Flowers* White with five petals, pink on the underside. *Fruit* Almost spherical, red turning black as it ripens.

Flowering	Dry season.
Habitat	Along river banks and damp places; area 1.
Distribution	Old World Tropics.
Note	Fruit is edible.

NAME *Ceropegia meyeri*

Climbing perennial herb. *Leaves* Opposite, oval with wavy margins and indented veins. *Flowers* White, tubular with a dark green cage-like structure at the top.

Flowering	Rains.
Habitat	Damp, shady places; area 1.
Distribution	Old World tropics.

NAME *Diplorhynchus condylocarpon*

Shrub or small deciduous tree with drooping branches, bark fissured; up to 6 m. *Leaves* Opposite, elliptic, shiny with pale yellow midrib, margins entire. *Flowers* White, in loose clusters near the end of branches, fragrant. *Fruit* Paired boat-shaped pods with a beak.

Flowering Early rains.

Habitat Open woodland and stony hills; area 6.

Distribution Tropical and southern Africa.

NAME *Orthanthera jasminiflora*

A prostrate herb; spreading up to 3 m. *Leaves* Opposite, lance-shaped with entire margins. *Flowers* Groups of cream/pale yellow flowers with five petals.

Flowering Early rains.

Habitat Sandy places; area 9.

Distribution Central and southern Africa.

Note Eaten as a vegetarian meat.

Apocynaceae (OLEANDER AND MILKWEED FAMILY)

NAME *Tacazzea apiculata*

Woody climber with milky latex. *Leaves* Opposite, oval, margins entire and apex pointed. *Flowers* Masses of creamy white flowers with reflexed petals and protruding stamens, growing in the leaf axils. *Fruit* Paired capsules.

Flowering	Rains.
Habitat	Flood plains and damp places; areas 8 and 9.
Distribution	Common in tropical Africa.

Araceae (AROID AND DUCKWEED FAMILY)

NAME *Pistia stratiotes*

A floating perennial often forming dense mats; up to 20 cm. *Leaves* A rosette, thick dull green, ridged, hairy. *Flowers* White, inconspicuous among leaves, single female below and a whorl of male flowers above. *Fruit* A green berry.

Flowering	All year.
Habitat	Lakes, dams and rivers; area 9.
Distribution	Pan-tropical invasive weed.
Note	Floating mats of *Pistia* reduce light and interfere with oxygen exchange, thus killing fish and other aquatic plants.

Asparagaceae (ASPARAGUS FAMILY)

NAME *Asparagus laricinus*

Spiny shrub, erect or climbing; up to 3 m. *Leaves* Needle-like whorls at intervals along the branches. *Flowers* White, small, borne along the branches. *Fruit* Green berry turning red as it matures, with one seed.

Flowering	Early rains.
Habitat	Along rivers, common creeping over shrubs; area 1.
Distribution	Central and southern Africa.

Asphodelaceae (ALOE FAMILY)

NAME *Aloe chabaudii*

Stemless with numerous suckers; up to 1 m. *Leaves* Erect rosettes, may curve outwards, greyish with well-spaced firm white teeth. *Flowers* Red, swollen at the base, on spreading branched stems.

Flowering	Early dry season.
Habitat	Dry, bare rocky hillsides; area 2.
Distribution	Throughout southern Africa.
Note	Intolerant of frost and grass fires.

NAME *Aloe cryptopoda*

Generally single plants; up to 1.7 m. *Leaves* A rosette of long, greyish leaves curving inwards. *Flowers* The inflorescence branches into 3-8 spikes with straight red flowers.

Flowering	Dry season.
Habitat	Dry rocky hillsides; area 1.
Distribution	Throughout southern Africa.
Note	First described by John Kirk on Livingstone's expedition in 1859. Intolerant of grass fires.

NAME *Aloe zebrina*

Often form dense groups; up to 2.5 m. *Leaves* Small rosettes, triangular, dullish grey with white markings. *Flowers* Dull red or orange with marked swelling at the base. *Fruit* A large capsule.

Flowering	Early dry season.
Habitat	Open grassland and waste ground; area 9.
Distribution	Widely distributed in southern and Central Africa.

NAME *Acanthospermum hispidum*

Erect, much branched annual with hairy stems; up to 1.3 m. *Leaves* Opposite, oval, hairy with either small serrations or entire margins. *Flowers* Yellow, small with enlarged spiny bracts clustered in heads. *Fruit* Flattened, triangular covered in stiff hooked hairs with paired spines at the top.

Flowering	Rains.
Habitat	Common weed; area 9.
istribution	Native to tropical America, now pan-tropical.

NAME *Ageratum conyzoides*

An erect annual with hairy stems; up to 1 m. *Leaves* Opposite, heart-shaped with serrated margins. *Flowers* Terminal clusters of white or mauve flowers. *Fruit* Has pointed bristles, dispersed by wind.

Flowering	Rains.
Habitat	Mainly in deciduous woodland; areas 7 and 9.
Distribution	Native to the Caribbean, now pan-tropical.
Note	Numerous reported medicinal uses, including as a cure for epilepsy. Has been shown to have anti-bacterial properties

Asteraceae (DAISY AND SUNFLOWER FAMILY)

NAME *Aspilia kotschyi*

An erect, hairy herb; up to 1.2 m. *Leaves* Opposite, hairy, narrow with entire margins and three conspicuous veins. *Flowers* Deep red, with a very hairy calyx.

Flowering Rains.

Habitat Open grassland on well-drained soils; area 9.

Distribution Tropical Africa.

Note There is a white variety which grows on poorly drained alkaline soils.

NAME *Bidens pilosa*

Erect annual, branching above; up to 1 m. *Leaves* Opposite, trifoliate with serrated margins. *Flowers* White petals with yellow centres, often absent, or may remain closed. *Fruit* Black, barbed.

Flowering Most of the year.

Habitat Open sunny areas, often near water and as a weed of cultivation; area 9.

Distribution Pan-tropical weed.

Note Is highly invasive, one plant can produce up to 6000 fruits a year. *B. biternata*, with yellow flowers and lobed leaves, is also common.

Asteraceae (DAISY AND SUNFLOWER FAMILY)

NAME *Blumea decurrens*

An erect herb covered in soft white hair; up to 1 m. *Leaves* Alternate, grey, clasping the stem. *Flowers* Yellow, borne in the leaf axils.

Flowering Dry season.

Habitat Disturbed ground; **area 9.**

Distribution Weed in Central and southern Africa, often growing in large stands.

Note Has some anti-bacterial and anti-fungal properties.

NAME *Calostephane divaricata*

Open, branched annual with slender stems; up to 1 m. *Leaves* Alternate, with finely serrated or entire margins. *Flowers* Terminal, deep yellow.

Flowering Early dry season.

Habitat Open grassland; **area 9.**

Distribution Central and southern Africa.

NAME *Conyza aegyptiaca*

Erect annual, aromatic, with hairy stems; up to 1.5 m. *Leaves* Alternate, greyish/green, variable in shape, may be deeply lobed. *Flowers* Yellow, on long stalks.

Flowering Early rains.

Habitat Open woodland and disturbed ground; area 9.

Distribution Old World tropics.

NAME *Conyza bonariensis*

A robust erect annual with hairy stems; up to 1 m. *Leaves* Alternate, grey/green, linear with small hairs, smooth or finely serrated margins. *Flowers* Terminal, pale yellow flower heads. *Fruit* Oblong with pale brown or pink bristles.

Flowering Most of the year.

Habitat Disturbed areas and waste ground; areas 1 and 9.

Distribution A pan-tropical weed, probably native to Central America.

Note Herbicide resistant

Asteraceae (DAISY AND SUNFLOWER FAMILY)

NAME *Denekia capensis*

Slender, erect annual, branched from the base; up to 80 cm. *Leaves* Alternate, narrow with broadly serrated margins. *Flowers* Groups of pale mauve to purple flower heads on stalks.

Flowering Late rains.

Habitat Locally common in damp open places, prefers clay soil; area 6a.

Distribution Angola, Namibia and southern Africa.

Note Can be confused with *Erlangea misera*

NAME *Ethulia conyzoides*

An erect annual, the whole plant is aromatic; 25–180 cm tall. *Leaves* Alternate, with irregularly serrated margins. *Flowers* Groups of mauve flower heads are terminal. *Fruit* Ribbed.

Flowering Rains.

Habitat In moist areas along rivers; area 9.

Distribution Mainly tropical Africa, also found in tropical Asia.

Note Has some molluscicide activity.

NAME *Flaveria trinervia*

Branching annual with reddish stems; up to 80 cm. *Leaves* Opposite, lance-shaped, with marked veins and serrated margins. *Flowers* Yellow in dense clusters., borne in leaf axils and terminally.

Flowering	Early rains.
Habitat	Disturbed and waste ground; area 9.
Distribution	Pan-tropical weed.

NAME *Grangea maderaspatana*

Loosely branched annual with hairy stems; up to 50 cm. *Leaves* Alternate, hairy, with lobed margins. *Flowers* Terminal yellow heads.

Flowering	Early rains.
Habitat	Damp areas along rivers; area 8.
Distribution	Old World Tropics.

NAME *Helichrysum argyrosphaerum*

Prostrate or low-growing branching annual; up to 50 cm. *Leaves* Alternate with entire margins, covered in white woolly hair. *Flowers* Solitary at the end of branches, silver with pink centre, turning yellow then brown with age.

Flowering Most of the year.

Habitat Sandy, disturbed ground; areas 6 and 9.

Distribution Central and southern Africa.

Note Poisonous to mammals.

NAME *Nidorella resedifolia*

Erect perennial: up to 75 cm. *Leaves* Alternate, with small terminal points and entire margins. *Flowers* Clusters of bright yellow flower heads. *Fruit* Slightly flattened.

Flowering Rains.

Habitat Damp, shady areas on sand or clay, also on disturbed ground; area 9.

Distribution Central and southern Africa.

Note Leaves edible.

NAME *Pegolettia senegalensis*

Erect branching herb with hairy stems; up to 60 cm. *Leaves* Alternate, linear with notched margins. *Flowers* Yellow heads, terminal.

Flowering Late rains.

Habitat Sandy open woodland; area 5.

Distribution Old World Tropics.

NAME *Sclerocarpus africanus*

Annual herb, hairy; up to 1.2 m. *Leaves* Opposite below, alternate above, oval with serrated edges, hairy. *Flowers* Terminal, above five leafy bracts, ovary forms a large bulge with a few yellow ray florets above. *Fruit* Brown/black, ovoid.

Flowering Rains.

Habitat Disturbed ground; area 1.

Distribution Tropical Africa and east to India.

NAME *Sonchus asper*

An erect annual with milky sap; up to 1 m. *Leaves* Alternate, upper leaves smaller, acute tip, margins wavy and deeply serrated. *Flowers* Terminal yellow flower heads.

Flowering	Rains.
Habitat	Disturbed ground; area 9.
Distribution	Pan-tropical weed.
Note	May be confused with *Sonchus oleraceus* which has less acute leaf tips.

NAME *Tithonia rotundifolia*

Tall annual, stems may be purplish; up to 2 m. *Leaves* Alternate, ovate with finely serrated margins and soft hair on the underside. *Flowers* Bright orange, erect, up to 7 cm in diameter.

Flowering	Rains.
Habitat	Along roadsides and in damp, disturbed ground; area 9.
Distribution	Native to Mexico, now widely distributed in the tropics.
Note	Named for Tithonus, a legendary Trojan.

NAME *Tridax procumbens*

A sprawling, hairy perennial; up to 40 cm. *Leaves* Opposite, hairy, with irregular lobes and coarsely serrated margins. *Flowers* White with yellow centres, borne on long erect hairy stalks. *Fruit* Black, hairy with bristles on top.

Flowering	Most of the year.
Habitat	Along roadsides and disturbed ground; area 9.
Distribution	Pan-tropical weed, originally native to C and tropical S America. Very common.

NAME *Vernonia adoensis*

A soft-wooded shrub; up to 2 m. *Leaves* Alternate, broadly oval with strongly undulating margins. *Flowers* Several large flower-heads on stalks, mauve/white surrounded by several loose bracts.

Flowering	Late rains.
Habitat	Open woodland and wooded grassland, often near streams; area 6a.
Distribution	Tropical Africa.
Note	There are 14 species of *Vernonia* in the area.

NAME *Vernonia glabra*

An erect annual; up to 1.0 m.
Leaves Alternate, with entire margins.
Flowers Deep purple flower-heads.
Fruit Hairy.

Flowering	Rains.
Habitat	Deciduous woods and grassland; area 9.
Distribution	Old World tropics.
Note	Used to cure burns and surface wounds.

NAME *Xanthium strumarium*

An erect annual with a short hairy stem; up to 1.0 m. *Leaves* Alternate, broadly triangular with three shallow lobes and widely serrated margins. *Flowers* White or green, covered in hooked bristles, grouped in the leaf axils or terminally. *Fruit* with two hooked bristles.

Flowering	Early dry season.
Habitat	Disturbed areas and river banks, requires full sun; area 7.
Distribution	Native to America, now a pan-tropical weed.
Note	Leaves produce tannin and a yellow dye. The dried plant repels weevils in stored wheat. Contains glycoside and, in spite of its numerous medicinal uses, is probably poisonous, particularly to livestock.

Bignoniaceae (TRUMPET VINE, JACARANDA AND CATALPA FAMILY)

NAME *Markhamia zanzibarica*

A large shrub with slender branches; up to 3 m. *Leaves* Opposite, compound with 2-4 pairs of leaflets and a terminal leaflet, margins smooth. *Flowers* Trumpet-shaped, yellow outside with maroon flecks, maroon inside. *Fruit* Slender, dark brown capsule with numerous winged seeds.

Flowering Early rains.

Habitat The edges of rivers and rocky outcrops; areas 8 and 9.

Distribution Tropical Africa.

Boraginaceae (BORAGE AND FORGET-ME-NOT FAMILY)

NAME *Cordia pilosissima*

Scrambling shrub with rough grey bark, hairy throughout; up to 10 m. *Leaves* Alternate, ovate with entire margins. *Flowers* Creamy with yellow hairs. *Fruit* Fleshy, orange when ripe.

Flowering Early rains.

Habitat Deciduous thickets, often close to water; area 8.

Distribution Central Africa.

Boraginaceae (BORAGE AND FORGET-ME-NOT FAMILY)

NAME *Heliotropium baclei*

An erect hairy perennial; up to 80 cm. *Leaves* Alternate, oval with entire margins. *Flowers* Yellow solitary with five fused petals. *Fruit* Round with four nutlets.

Flowering	Rains.
Habitat	Open, dry river banks; area 8.
Distribution	Tropical Africa.
Note	*Heliotropium* is from the Greek meaning turning to the sun.

NAME *Heliotropium ovalifolium*

Perennial herb, much branched; up to 85 cm. *Leaves* Alternate, with entire margins. *Flowers* Small white, arranged in two rows on a curving spike. *Fruit* Round splitting into four nutlets.

Flowering	Most of the year.
Habitat	Open, sunny places; areas 1, 8 and 9.
Distribution	Old World tropics.
Note	Topical medicinal uses in some African countries. Reported to be poisonous to horses in Australia.

Boraginaceae (BORAGE AND FORGET-ME-NOT FAMILY)

NAME *Trichodesma zeylanicum*

A very hairy erect annual with ascending branches; up to 1.5 m. *Leaves* Opposite, with entire margins. *Flowers* Pale blue/white, terminal, face downwards. *Fruit* Four grey/brown nutlets.

Flowering Early dry season.

Habitat Disturbed ground, a weed of cultivation; area 9.

Distribution Old World tropics.

Note Several, unrelated, medicinal properties are reported. Poisonous to birds.

Byttneriaceae (CHOCOLATE FAMILY)

NAME *Waltheria indica*

Shrub with soft, grey hairs; up to 2 m. *Leaves* Alternate, with serrated margins and conspicuous veins. *Flowers* Clusters of small, fragrant, yellow flowers borne in the leaf axils. *Fruit* Dry capsule with one, tiny, black seed.

Flowering Rains.

Habitat Dry grassland; area 6.

Distribution Pan-tropical, probably native to tropical America.

Note Used in cosmetics to lighten the skin. Good animal forage containing up to 6% protein, in the wet season.

NAME *Lobelia kirkii*

Prostrate or scrambling perennial herb; up to 40 cm. *Leaves* Basal leaves are oval with serrated margins, upper leaves are alternate, linear with small serrations. *Flowers* Blue with lobed corolla, in loose sprays. *Fruit* Capsule with numerous small seeds.

Flowering Mid-late rains.

Habitat Open spaces within the spray zone; area 1.

Distribution Only recorded in the Victoria Falls area.

Note Flowers attract butterflies.

NAME *Wahlenbergia denticulata*

Erect herb; up to 90 cm. *Leaves* Alternate, narrow with undulating margins. *Flowers* Blue with a paler centre and five open petals. *Fruit* Capsule with three lobes.

Flowering Rains.

Habitat Disturbed areas, rocky or seasonally damp ground; areas 1, 7 and 8.

Distribution Central and southern Africa.

Capparaceae (CAPER FAMILY)

NAME *Capparis tomentosa*

A robust woody climber or shrub with paired hooked thorns; up to 5 m. *Leaves* Alternate, grey/green, hairy with entire margins. *Flowers* Yellow/green with a mass of untidy long white/pink stamens. *Seeds* Embedded in red pulp.

Flowering	Dry season.
Habitat	Deciduous woodland; areas 7 and 8.
Distribution	N.E. and southern Africa.
Note	Toxic, causes jaundice.

NAME *Maerua juncea*

Woody climber or scraggy shrub with green stems; up to 8 m. *Leaves* Alternate, oval with entire margins. *Flowers* White/green, small petals with conspicuous long stamens, clustered at the end of the branches.

Flowering	Dry season.
Habitat	In hot dry acacia woodland; area 9.
Distribution	Tropical Africa.
Note	The larval food plant of several Pierid (white) butterflies; adults attracted to flowers.

Cleomaceae (SPIDER FLOWER AND AFRICAN CABBAGE FAMILY)

NAME *Cleome gynandra*

An erect annual with sticky stems; up to 60 cm. *Leaves* Palmately compound with three to five egg-shaped leaflets arising from the stem. *Flowers* White sprays with prominent stamens, terminal. *Fruit* Long, pod-like with brown seeds.

Flowering Early dry season.

Habitat Common weed in disturbed ground; area 9.

Distribution Pan-tropical.

Note Traditionally used to cure rheumatoid arthritis. Is host to tobacco aphids in winter which can cause destruction to crops. Leaves are a nutritious and commonly eaten vegetable, 'suntha'.

NAME *Cleome hirta*

Erect annual with a hairy stem; up to 1.5 m. *Leaves* Alternate, compound with 5-9 narrow leaflets, sticky, margins uneven, aromatic. *Flowers* Mauve alternating along the erect spike. *Fruit* Capsule with ridged seeds.

Flowering Rains.

Habitat Open woodland where rainfall is less than 750 mm per annum; areas 5, 6, 7, and 9.

Distribution Throughout southern, Central and East Africa.

Note *Cleome* is the ancient name for a mustard-like plant.

Colchicaceae (COLCHICUM FAMILY)

NAME *Gloriosa superba*

A scrambling perennial, up to 1.5 m. *Leaves* Usually alternate, with entire margins. Leaf tip ends in a coiling tendril. *Flowers* Orange/red with six reflexed tepals. Six long stalked stamens spread horizontally below the flower. *Fruit* Leathery capsule with bright red seeds.

Flowering Rains.

Habitat Widespread in grass, woodland and the margins of thickets; areas 5, 8 and 9.

Distribution Old World tropics.

Note The national flower of Zimbabwe and the state flower of Tamil Nadu in India. A treatment for gout since the 1st Century. Now licensed as an anti-inflammatory in small doses. Tubers contain the poison colchicine.

NAME *Combretum hereroense*

A small, deciduous tree or shrub, with dark grey, rough bark; up to 9 m. *Leaves* Opposite, crowded on the branches, with entire margins, velvety below. *Flowers* Cream/white or pale yellow borne on spikes, pleasant scent. *Fruit* Red/brown with four wings, remains on the shrub for several months.

Flowering Late dry season.

Habitat Loamy sand and grassland, often on termite mounds or around pans; areas 6, 7 and 9.

Distribution East, Central and southern Africa.

Note Ash from dried bark is made into a paste to straighten hair. Wood used for walking sticks and building supports. Browsed by animals.

NAME *Combretum mossambicense*

A deciduous, scrambling shrub or creeper covered in soft hair; up to 2 m. *Leaves* Opposite, hairy when young, elliptic with entire margins. *Flowers* Clusters of fragrant, pink or white flowers appear before the leaves. *Fruit* Four-or five-winged pinkish fruit ripen to grey/brown in summer.

Flowering Late dry season.

Habitat Open grassland and rocky outcrops; areas 5, 7 and 9.

Distribution Dry areas of East, Central and southern Africa.

Note Larvae of the Striped Policeman and Guinea-fowl butterflies feed on the leaves. The flowers draw insects, in turn attracting insect-eating birds such as the Brown-hooded Kingfisher.

Commelinaceae (SPIDERWORT FAMILY)

NAME *Commelina forskaolii*

Sprawling, slightly succulent perennial; up to 30 cm. *Leaves* Alternate with entire margins and parallel veins. *Flowers* Bright blue with two free petals above and one below. *Fruit* Green capsule with white hairs.

Flowering	Most of the year.
Habitat	Damp areas and woodlands; area 9.
Distribution	Tropical Africa, Arabia and India.
Note	There are 10 species of *Commelina* in the area.

NAME *Cyanotis longifolia*

Perennial herb, hairy; up to 90 cm. *Leaves* Alternate, hairy, lance-shaped with entire margins. *Flowers* Deep blue with conspicuous yellow anthers, borne between bracts in the leaf axils.

Flowering	Rains.
Habitat	Savannah and grassland; area 1.
Distribution	East, Central and southern Africa.
Note	Able to accumulate cobalt from the soil.

Connaraceae (ZEBRA WOOD FAMILY)

NAME *Rourea orientalis*

A scrambling shrub; up to 3 m.
Leaves Alternate, compound with 8-16 pairs of leaflets and a terminal leaflet, folded downwards. *Flowers* Loose sprays of white/cream flowers appear just before the leaves, lemon-scented. *Fruit* A green leathery capsule splitting along one seam; seed enclosed in a bright red coating.

Flowering Late dry season.
Habitat Hot, dry areas along rivers; areas 1 and 8.
Distribution Central Africa and Madagascar.

Convolvulaceae (MORNING GLORY AND BINDWEED FAMILY)

NAME *Astripomoea lachnosperma*

Erect, hairy annual; up to 120 cm.
Leaves Alternate, with entire margins below and wavy towards the tip. *Flowers* White funnel-shaped with a purple centre, crowded in the leaf axils. *Fruit* Ovoid, smooth capsule.

Flowering Late rains.
Habitat Sandy soils, grassland and open forest; area 9.
Distribution Widely distributed throughout Africa.

NAME *Convolvulus sagittatus*

A prostrate or twining perennial; up to 60 cm. *Leaves* Narrow, arrow-shaped with entire margins. *Flowers* White or pale pink, axillary either solitary or paired.

Flowering	Rains.
Habitat	Open sunny places; areas 6 and 9.
Distribution	Throughout Africa.

NAME *Evolvulus alsinoides*

Very slender erect plant; up to 80 cm. *Leaves* Alternate, narrow with entire margins, hairy. *Flowers* Bright blue, solitary, arising in the leaf axils. *Fruit* A round capsule.

Flowering	Rains.
Habitat	Sunny and damp places; area 1.
Distribution	Pan-tropical.
Note	Widely used in India to alleviate nervous conditions.

NAME *Ipomoea spp*

Annual or perennial lianas, shrubs or small trees. *Leaves* Alternate, often palmate with 3-5 lobes. *Flowers* Solitary, in leaf axils usually brightly coloured, trumpet-shaped. *Fruit* A capsule.

Note Widely used as ornamentals. Many are food crops e.g. *Ipomoea batatas* (Sweet potato).

NAME *Ipomoea aquatica*

Has hollow stems allowing it to float. Cultivated as a vegetable in SE Asia.

NAME *Ipomoea dichroa*

Prostrate annual. Found in dry sandy areas.

NAME *Ipomoea eriocarpa*

Prostrate annual in dry sandy areas.

NAME *Ipomoea mauritiana*

Climbing perennial in damp places.

Convolvulaceae (MORNING GLORY AND BINDWEED FAMILY)

NAME *Ipomoea pes-tigridis*

A sprawling annual; up to 80 cm.
On sandy river banks.

NAME *Ipomoea rubens*

A bushy climber; up to 2 m. Always near water.

NAME *Ipomoea shirambensis*

A perennial woody climber; up to 3 m. Leaves appear after flowering Common in dry, open woodland.

NAME *Ipomoea tricolor*

A perennial vine; up to 3 m. Escaped ornamental, found near habitations.

Convolvulaceae (MORNING GLORY AND BINDWEED FAMILY)

NAME *Turbina holubii*

Perennial shrub, deciduous, much branched from the base, hairy; up to 2.5 m. *Leaves* Alternate, covered in fine silvery hair, ovate, indented at the base with entire margins. *Flowers* Trumpet-shaped, varying in colour from pink to mauve with a darker centre, borne in the leaf axils. *Fruit* Leathery pods with one yellowish seed.

Flowering Rains.

Habitat Wooded grassland; area 5.

Distribution Central and southern Africa.

Crassulaceae (STONECROP AND HOUSELEEK FAMILY)

NAME *Kalanchoe lanceolata*

Succulent, erect annual covered in sticky glands; up to 1.5 m. *Leaves* Opposite, oblong with entire or wavy margins. *Flowers* Orange with four petals, in groups borne in the leaf axils. *Fruit* Small seeds with raised longitudinal lines.

Flowering Early dry season.

Habitat All types of woodland and rocky outcrops; area 9.

Distribution Common in tropical Africa, recorded in Yemen and India.

Note Attracts sunbirds.

NAME *Coccinia adoensis*

Perennial with annual climbing stems; up to 3 m. *Leaves* Extremely variable, broadly triangular, may be lobed, with finely serrated margins. *Flower* Yellow, male and female similar but on different plants. *Fruit* Red/yellow, ellipsoid, smooth.

Flowering	Early rains.
Habitat	Savannah; area 5.
Distribution	Tropical Africa.
Note	Leaves eaten as a vegetable.

NAME *Momordica balsamina*

Scrambling annual from woody rootstock. *Leaves* Almost circular in outline with 5 -7 deep lobes. *Flowers* Pale yellow often with green veins, solitary. *Fruit* Ovoid with warty skin, red/orange when ripe.

Flowering	Rains.
Habitat	Open sandy places; area 9.
Distribution	Native to tropical Africa, now pan-tropical.
Note	Can be confused with *Momordica charantia*. Has numerous medicinal uses throughout Africa.

NAME *Cyperus papyrus*

A stately herbaceous perennial forming grass-like clumps with triangular green stems; up to 5 m. *Leaves* Reduced, appears to be leafless. *Flowers* Greenish brown flowers cluster at the end of the rays that form a feather-like head.

Flowering Late rains, early dry season.

Habitat Along the edges of water courses; area 8.

Distribution African water courses and parts of the Mediterranean. An escaped ornamental in other parts of the tropics and sub-tropics.

Note Used extensively by the ancient Egyptians, for paper, boats, etc.

NAME *Kyllinga alba*

A densely tufted perennial with aromatic roots; up to 40 cm. *Leaves* Basal, linear with entire margins, up to 26 cm. *Flowers* Spherical white heads of winged spikelets.

Flowering Rains.

Habitat Open grassland; areas 1, 5, 7, 8 and 9.

Distribution Tropical Africa.

Dioscoreaceae (DIOSCOREA AND YAM FAMILY)

NAME *Dioscorea dumetorum*

Twining perennial with prickly annual stems; climbing up to 10 m. *Leaves* Alternate, compound with three leaflets and entire margins. *Flowers* Brown, hairy, on spreading, cylindrical spikelets. *Fruit* Brown, oblong, 3-winged capsule.

Flowering Rains.

Habitat Well drained shady places; area 1.

Distribution Africa, south of the Sahara.

Note Used in traditional medicine to treat schistosomiasis, diabetes and as a topical anaesthetic.

Dracaenaceae (DRAGON TREE FAMILY)

NAME *Sansevieria pearsonii*

Dark green and spiny, forming dense clumps; up to 80 cm. *Leaves* 7-8 in two ranks, erect, rigid, cylindrical, tapering to a hard spine-like tip, with several shallow grooves on sides and back. *Flowers* A long spike-like raceme of pale pink/brown flowers, opening at night. *Fruit* A green berry turning orange as it matures.

Flowering Dry season.

Habitat Dry scrubland; areas 1 and 8.

Distribution Central and southern Africa.

Note Produces fibre used for weaving.

NAME *Acalypha fimbriata*

Erect annual; up to 75 cm. *Leaves* Alternate, with long petioles, margins finely serrated. *Flowers* White on small erect spikes between bracts in the leaf axils.

Flowering Rains.

Habitat Damp, shady places; area 1.

Distribution Widespread in tropical Africa.

Note Leaves have medicinal properties, including as a cure for syphilis.

NAME *Acalypha ornata*

Erect shrub; up to 3 m. *Leaves* Alternate, margins finely serrated. *Flowers* Erect spikes with red female and small, white male flowers within bracts. *Fruit* A small three-lobed capsule splitting to release round grey/purple seeds.

Flowering Rains.

Habitat Damp, shady places; common in the rainforest area of the Falls; area 1.

Distribution Tropical Africa.

Note Several medicinal uses are reported, including as a cure for leprosy.

NAME *Croton gratissimus*

A deciduous shrub or small tree, with pendulous branches; up to 10 m. *Leaves* Alternate, dark green above and silver below, with smooth margins. *Flowers* Cream/yellow on spikes, appear before the leaves, buds remain on branches throughout the dry season. *Fruit* A three-lobed capsule, green, turning yellow.

Flowering Early rains.
Habitat Dry rocky hillsides; area 6.
Distribution Tropical Africa.

NAME *Euphorbia griseola*

Succulent shrub, much branched, a spine shield forms a horny ridge along the angles; up to 3.5 m. *Leaves* Absent. *Flowers* Yellow turning red along the ridges. *Fruit* A three-lobed capsule.

Flowering Dry season.
Habitat Among granite rocks in open woodland; area 3.
Distribution Central Africa.
Note The name Euphorbia is after the Greek physician to King Juba II of Numidia (50 BC – 23 AD).

NAME *Euphorbia heterophylla*

A semi-woody, erect annual with milky sap; up to 1 m. *Leaves* Alternate, with smooth margins. *Flowers* Greenish/pink, densely clustered in leaf nodes or terminally. Large bracts may be tinged red or have white splashes.

Fruit	A three-lobed capsule with conical seeds.
Flowering	Most of the year.
Habitat	Disturbed and waste ground; area 1.
Distribution	Common in the tropics. Native to the West Indies.
Note	The whole plant is poisonous.

NAME *Euphorbia oatesii*

Decumbent perennial herb with annual stems from a bulbous tuber; up to 20 cm. *Leaves* Alternate, linear with smooth margins. *Flowers* Pink, with feathery stamens, terminal. *Fruit* A three-lobed capsule.

Flowering	Early rains.
Habitat	Stony ground in open grassland; area 6a.
Distribution	Only reported from the Zambezi region.

NAME *Ricinus communis*

Erect perennial with woody base and a single hollow stem; up to 2 m. *Leaves* Alternate, glossy green, palmately lobed with serrated margins. *Flowers* Red female above and white male flowers below, on a terminal spike. *Fruit* Round, white capsules with dark spines. Seeds mottled brown.

Flowering Most of the year, if near water.

Habitat Disturbed and waste ground; area 9.

Distribution Native to tropical NE Africa and the Middle East, now pan-tropical.

Note Cultivated commercially for its oil (from the seeds), which produce castor oil and have numerous medicinal and industrial purposes. The poison ricin (in the seeds) is water-soluble and not present in the oil.

NAME *Tragia okanyua*

A climbing perennial herb, hairy; stems up to 3 m. *Leaves* Alternate, with serrated margins, may be three-lobed, hairy. *Flowers* Round white heads with green stripes. *Fruit* Green/yellow with three lobes. Seeds small, greyish with brown flecks.

Flowering Rains.

Habitat Open grassland and thicket margins; areas 1 and 9.

Distribution Tropical Africa.

Note Hairs on the leaves and inflorescence can cause skin irritation.

NAME *Albuca abyssinica*

Perennial herb from a bulb with an annul stem; up to 1 m. *Leaves* A basal rosette, linear with smooth margins. *Flowers* Up to 100 yellow flowers with a green midrib, on a tall spike.

Flowering	Early rains.
Habitat	Disturbed ground, along roadsides and open places; area 9.
Distribution	Throughout Africa and into Arabia.

NAME *Dipcadi platyphyllum*

Perennial herb from a bulb; up to 20 cm. *Leaves* One single leaf, lance-shaped with wavy margins. *Flowers* Green/yellow, up to 30 flowers on a spike.

Flowering	Early rains.
Habitat	Common in open grassland; areas 2 and 7.
Distribution	Tropical and southern Africa.

NAME *Ledebouria zambesiaca*

Perennial herb from a bulb; up to 30 cm. *Leaves* Basal rosette, may have maroon spots. *Flowers* Mauve with green sepals, up to 100 flowers clustered on a spike.

Flowering	Early rains.
Habitat	Open woodland, very common; areas 2,5,6,7 and 9.
Distribution	Tropical Africa.

NAME *Urginea altissima*

Perennial herb; up to 2 m. *Leaves* Small basal rosette of greyish/green leaves. *Flowers* White with green midrib, up to 100 on a long spike.

Flowering	Early rains.
Habitat	Open woodland; areas 1 and 9.
Distribution	Native to southern Africa.
Note	Used to cure skin problems and rheumatism.

NAME *Gladiolus dalenii*

Annual; up to 60 cm. *Leaves* Basal, long, narrow with entire margins. *Flowers* Up to 14 large yellow flowers borne on an erect spike. *Fruit* An oblong capsule.

Flowering	Rains.
Habitat	In damp grass and open woodlands; area 1.
Distribution	Wetter areas of southern Africa and Madagascar.
Note	Common ornamental and root stock for many hybrids. Used medicinally to treat dysentery and constipation. Recognised as a cure for snake-bite. Corms can be boiled and eaten.

NAME *Lapeirousia erythrantha*

Erect annual with delicate branching stems; up to 60 cm. *Leaves* Linear, lower two longer than upper ones. *Flowers* Six mauve petals emerge from a long tubular corolla, opening in the evening and early morning.
Fruit Capsule.

Flowering	Rains.
Habitat	Locally damp sites; area 9.
Distribution	Sub-Saharan Africa.
Note	Corms are edible either raw or cooked.

Lamiaceae (MINT FAMILY)

NAME *Hemizygia bracteosa*

An erect perennial; up to 40 cm. *Leaves* Opposite, narrow with finely serrated margins. *Flowers* White or pale mauve, inconspicuous beneath white or purple bracts. *Fruit* Berry.

Flowering	Rains.
Habitat	Common in wooded grassland; areas 7 and 9.
Distribution	Tropical and southern Africa.

NAME *Hoslundia opposita*

Shrubby perennial with angled branches; up to 1.2m. *Leaves* Opposite, with very small serrations, has an unpleasant smell that repels bees. *Flowers* Small cream/green in clusters at the end of branches. *Fruit* Orange, fleshy, berry-like with a sweet taste.

Flowering	Rains.
Habitat	Wetter areas around the Falls; area 1.
Distribution	Common throughout Africa.
Note	Fruit is edible and attracts birds. Roots contain sedatives used in traditional medicine.

NAME *Kalaharia uncinata*

Dense shrubby perennial, with recurved spines; up to 1.2 m. *Leaves* Opposite, elliptic, hairy with entire margins. *Flowers* Showy red with paler centres, petals reflexed, prominent stamens. *Fruit* Ovoid capsule, with 2-4 lobes, shiny, black.

Flowering End of the dry season.

Habitat Open and disturbed ground, often on Kalahari sand; area 6.

Distribution Central Africa.

NAME *Leonotis nepetifolia*

An erect, loosely branched annual with square stems; up to 2.5 m. *Leaves* Opposite, narrow with coarsely serrated margins. *Flowers* Orange or red, furry, in clusters at intervals along the stem.

Flowering Early dry season.

Habitat Disturbed ground; areas 6 and 9.

Distribution Native to tropical Africa and southern India, now pan-tropical.

Note Has a number of traditional medicinal uses, dried leaves are commonly used as a mild narcotic. Attracts hummingbirds.

NAME *Leucas martinicensis*

An erect hairy annual with square stems, sometimes branched; up to 1.2 m. *Leaves* Opposite, narrow with coarsely serrated margins. *Flowers* White, emerge from a hairy, ball-like calyx that arises at the axils of the upper pair of leaves. *Fruit* Four dark brown, oblong nutlets.

Flowering Early dry season.

Habitat Open grassland; area 9.

Distribution Pan-tropical weed of cultivation.

Note Plant has a minty smell and, in China, is burned to repel mosquitoes.

NAME *Ocimum sp.*

Woody perennial with square stems, aromatic; up to 80 cm. *Leaves* Opposite, appearing trifoliate with entire margins. *Flowers* Pale pink, two-lipped, with very long stamens, borne in whorls along an erect spike. *Fruit* Three to four oblong nutlets.

Flowering Most of the year.

Habitat Disturbed ground; areas 5 and 7.

Distribution Old World tropics.

Lamiaceae (MINT FAMILY)

NAME *Vitex petersiana*

A shrub or small tree; up to 2.5 m. *Leaves* Opposite with five leaflets, oval with entire or irregularly notched margins. *Flowers* Blue, irregular lobes, fragrant. *Fruit* Blue/black when mature.

Flowering	Rains.
Habitat	In open Mopane woodland; areas 5 and 7.
Distribution	Central Africa.

Leguminosae (PEA, BEAN AND MIMOSA FAMILY)

NAME *Baphia massaiensis*

Erect shrub; up to 2 m. *Leaves* Alternate, simple, obovate with entire margins. *Flowers* White with yellow spot on the main petal, pea-like, in short sprays, jasmine-scented. *Fruit* Oblong, thin-woody pod, hairy.

Flowering	Late rains and early dry season.
Habitat	Woodland scrub and Kalahari sand; areas 5 and 6.
Distribution	Central Africa.

Leguminosae (PEA, BEAN AND MIMOSA FAMILY)

NAME *Bauhinia petersiana*

Perennial shrub; may scramble up to 2 m. *Leaves* Alternate, butterfly-shaped with two deep lobes, margins entire. *Flowers* White, tinged with pink, crinkled, untidy petals. *Fruit* Dark brown woody pods splitting to release flattened circular seeds.

Flowering	Rains.
Habitat	Disturbed areas, wooded grassland, often on Kalahari sand; all areas.
Distribution	Central and southern Africa.
Note	There are two sub-species of *Bauhinia petersiana*; *macrantha* with broader petals and shorter stamens, usually found on Kalahari sands; and *petersiana* which has narrow petals and protruding red stamens.

NAME *Chamaecrista absus*

A small annual with hairy, sticky stems; up to 1 m. *Leaves* Two pairs of opposite leaflets with entire margins. *Flowers* Yellow or orange with reddish veins, terminal. *Fruit* Pods glandular with flat brown seeds.

Flowering	Rains.
Habitat	Disturbed ground and open grassland; areas 7 and 9.
Distribution	Pan-tropical.
Note	A wide variety of medicinal uses reported from all parts of the plant. Seeds used to make necklaces.

NAME *Chamaecrista mimosoides*

A robust, erect, shrubby plant with a hollow stem; up to 1.2 m. *Leaves* Compound with 30-60 paired leaflets. *Flowers* Yellow, one to three in the leaf axils. *Fruit* A flat linear pod.

Flowering Rains.

Habitat Forest margins, grassland and disturbed ground, may grow in standing water; areas 5, 7 and 9.

Distribution Common in the Old World tropics.

NAME *Crotalaria laburnifolia*

Erect perennial; up to 1.2 m. *Leaves* Trifoliate, leaflets oval with entire margins. *Flowers* Yellow, on a spike, lower ones open first. *Fruit* Chunky pods containing black beans.

Flowering Rains.

Habitat Disturbed areas and open woodland; area 9.

Distribution Old World tropics and Australia.

Note Leaves and pods are edible.

NAME *Decorsea schlechteri*

Perennial twining herb, deciduous; climbs up to 3 m. *Leaves* Compound with a terminal leaflet. *Flowers* Appear before the leaves, pale mauve, bluish or sometimes pale pink, pea-like. *Fruit* Pod.

Flowering	Late dry season.
Habitat	Gravel and Kalahari sand; area 9.
Distribution	Central and southern Africa.

NAME *Dichrostachys cinerea*

Semi-deciduous, fast growing woody shrub with alternate thorns; up to 5 m. *Leaves* Compound with 5-10 paired leaflets each with 10-30 sub-leaflets. *Flowers* Resemble a Chinese lantern, upper, pink/mauve staminodes are sterile with yellow flowers below. *Fruit* Twisted, mustard-brown pods.

Flowering	Early rains.
Habitat	Sandy and disturbed places; areas 6 and 9.
Distribution	Native to Old World tropics; introduced to the West Indies.
Note	Seeds edible, fruit favoured by animals. Wood is slow-burning and non-toxic. Invasive, causes crop losses in the West Indies.

NAME *Indigofera flavicans*

Decumbent perennial, hairy throughout; up to 20 cm. *Leaves* Compound, on ascending stems. *Flowers* Pink, on an erect spike. *Fruit* Long, pale green pods.

Flowering	Rains.
Habitat	Grassland, areas 5 and 6.
Distribution	Old World tropics.
Note	There are 24 species of *Indigofera* in the region.

NAME *Indigofera setiflora*

Erect, much branched perennial; up to 80 cm. *Leaves* Alternate, compound with 2-4 pairs of leaflets arranged alternately and folding upwards along the stem. *Flowers* Pink, often with a white spot inside, axillary. *Fruit* Pods, densely covered in short hairs.

Flowering	Rains.
Habitat	Grassland and rocky outcrops; areas 7 and 9.
Distribution	Tropical and southern Africa.
Note	The name *Indigofera* means to produce the dye indigo.

NAME *Mimosa pigra*

Dense shrub with randomly distributed backward curving prickles; up to 3 m. *Leaves* Compound, bi-pinnate with 15 pairs of leaflets which fold when touched, or at night. *Flowers* Pink, grouped into globular heads, two heads borne on stalks in each axil. *Fruit* Groups of thick, hairy, curved pods.

Flowering Dry season.

Habitat Seasonally inundated grassland; areas 1 and 8.

Distribution Native to tropical America. Invasive weed in Africa, Asia, Pacific Islands and northern Australia.

NAME *Mundulea sericea*

Shrub, or small tree with corky bark; 0.5-7 m. *Leaves* Alternate, compound with 4-8 pairs of leaflets and a terminal leaflet, covered in silky hairs. *Flowers* Mauve or violet, pea-like, in clusters. *Fruit* Pods covered in silky brown hair, turning grey, remain on branches till next flowering.

Flowering Early rains.

Habitat Rocky grasslands; area 2.

Distribution Old World tropics and South Africa.

NAME *Rhynchosia minima*

A small, hairy, creeping plant; up to 20 cm. *Leaves* Compound with three oval/circular leaflets, margins entire. *Flowers* Yellow with dark veins on the largest petal, on a loose spike with the lower ones opening first. *Fruit* Oblong pods with greyish black seeds.

Flowering Rains.

Habitat Sandy disturbed ground; **area 5.**

Distribution Widely distributed in the tropics and semi-tropics.

NAME *Vigna luteola*

Annual or perennial trailing herb with hairy stems; stems up to 6 m. *Leaves* Alternate, trifoliate with entire margins. *Flowers* Yellow, pea-like, on stalks. *Fruit* Linear pods, pendent.

Flowering Rains.

Habitat Damp places and river banks; **area 1.**

Distribution Common throughout the tropics and sub-tropics.

Note Good choice for regenerating soil in wet areas.

NAME *Vigna parkeri*

A climbing or prostrate perennial, may form dense mats. *Leaves* Alternate, trifoliate with round or oval leaflets. *Flowers* Deep purple, borne in alternate pairs in leaf axils. *Fruit* Compressed oblong pods with grey/black mottled seeds.

Flowering Rains.

Habitat Grassland, open woodland and as a weed of cultivation; areas 1, 8 and 9.

Distribution Tropical Africa.

NAME *Zornia milneana*

Perennial herb with decumbent or prostrate stems; up to 75 cm. *Leaves* Whorls of four narrow leaves with glands dotted underneath, margins entire. *Flowers* Yellow, with red veining, terminal with bright green bracts. *Fruit* Pod with 4-7 constricted segments, covered in spiny bristles.

Flowering Rains.

Habitat Sandy soils, grassland and floodplains; area 5.

Distribution Central and southern Africa.

Loranthaceae (MISTLETOE FAMILY)

NAME *Tapinanthus sp.*

A semi-parasite growing on the branches of trees. Stems grey with numerous lenticels. *Leaves* Opposite, dull green with entire margins. *Flowers* Bright red erect stamens, white at the base. *Fruit* Oblong, red berry with persistent calyx.

Flowering Early dry season.

Habitat Large trees in Kalahari sand; area 9.

Distribution Central and southern Africa.

Note The sticky seed germinates after birds wipe it off their beaks onto a branch. The seeds are also chewed by children to make a glue that they then spread on branches to catch birds.

Lythraceae (POMEGRANATE AND HENNA FAMILY)

NAME *Ammannia auriculata*

An erect annual branching from the base; up to 60 cm. *Leaves* Opposite, narrow with entire margins. *Flowers* Bright pink with four petals, borne in the leaf axils. *Fruit* Dark red, berry-like.

Flowering Late rains.

Habitat Damp open areas and grassland; area 7.

Distribution Pan-tropical weed.

Note Herbicide resistant. Similar in appearance to *Nesaea ondongana.*

Malvaceae (COTTON MALLOW AND HOLLYHOCK FAMILY)

NAME *Abutilon angulatum*

A tall perennial covered in fine white down; up to 3 m. *Leaves* Alternate, heart-shaped with sparsely serrated margins. *Flowers* Single, yellow with protruding stamens. Open in the late afternoon. *Fruit* Seeds in densely packed sections.

Flowering Early dry season.

Habitat Damp, open places; area 8.

Distribution Common in drier parts of Africa and Madagascar.

NAME *Azanza garckeana*

A deciduous shrub or multi-stemmed small tree; up to 10 m. *Leaves* Alternate, circular in outline with 3 – 5 shallow lobes and a notch at the base, margins entire. *Flowers* Yellow or purplish with dark red centres. *Fruit* A capsule with woolly seeds.

Flowering Late rains.

Habitat Wide range, riverine, grassland and woodland; area 8.

Distribution East and southern tropical Africa.

Note An important food source; fruit high in protein and vitamin C, also contains calcium, phosphorus, sodium, magnesium, iron and potassium.

Malvaceae (COTTON MALLOW AND HOLLYHOCK FAMILY)

NAME *Hibiscus spp.*

There are 29 *Hibiscus* species in the Victoria Falls region.

Annual or perennial herbs or shrubs, often with hairy stems. *Leaves* alternate, with 3-5 shallow or deep lobes usually serrated. *Flowers* Solitary, showy with protruding staminal column, borne in the leaf axils. *Fruit* A capsule with hairy, kidney-shaped seeds.

Flowering	Rains.
Habitat	Damp and sandy areas.
Distribution	Common throughout the tropics, sub-tropics and temperate regions.
Note	Ornamental, often used as a hedge, medicinal and culinary uses e.g. as herbal infusions.

NAME *Hibiscus articulatus*

Grows up to 80 cm. Leaves deeply lobed. Flower white or yellow.

NAME *Hibiscus cannabinus*

Grows up to 3.5 m. Grown commercially for stem fibres which are used for rope, coarse cloth and paper.

Malvaceae (COTTON MALLOW AND HOLLYHOCK FAMILY)

NAME *Hibiscus mutatus*

Shrub up to 2.5 m. Small flowers open white, turn pink.

NAME *Hibiscus praeteritus*

Under-shrub, up to 2 m.

NAME *Hibiscus sidiformis*

Grows up to 1.0 m. Leaves deeply lobed, with entire margins.

NAME *Hibiscus surattensis*

Shrub, up to 2 m. Leaves deeply lobed with serrated margins.

Malvaceae (COTTON MALLOW AND HOLLYHOCK FAMILY)

NAME *Sida cordifolia*

Erect annual with hairy stems; up to 1 m. *Leaves* Alternate, heart-shaped with serrated margins. *Flowers* Solitary, yellow, towards the tips of branches, or terminal. *Fruit* Seeds smooth greyish/black.

Flowering Most of the year.

Habitat Weed of cultivation; areas 1 and 9.

Distribution Pan-tropical.

Note Numerous medicinal uses, especially in India, common ingredient of slimming tablets. Contains ephedrine.

NAME *Urena lobata*

Erect annual; up to 2.0 m. *Leaves* Alternate with 3-5 lobes, broadly round with serrated margins, covered in soft hair. *Flowers* Bright pink, solitary in the leaf axils. *Fruit* Flattened ball with five seeds.

Flowering Rains.

Habitat River banks; areas 1 and 8.

Distribution Pan-tropical, classified as a noxious weed in Florida.

Note In some countries, e.g. Brazil, it is cultivated for its fibre which is used to make sacks. Methanol extract from the roots shows broad anti-bacterial activity.

Menyanthaceae (BOG BEAN FAMILY)

NAME *Nymphoides indica*

Aquatic perennial with un-branched cylindrical stems. *Leaves* Circular with entire margins. *Flowers* White with yellow centre, five very hairy petals.

Flowering	Dry season.
Habitat	Still or slow moving water; area 8.
Distribution	Pan-tropical.

Molluginaceae (CARPET WEED FAMILY)

NAME *Glinus oppositifolius*

Prostrate annual; up to 50 cm. *Leaves* Opposite or in pseudo-whorls, with entire margins. *Flowers* White/green with five sepals, no petals. *Fruit* Oblong capsule with 5 valves.

Flowering	Rains.
Habitat	Damp, shady places; area 7.
Distribution	Old World tropics.

Nyctaginaceae (BOUGAINVILLEA AND FOUR O' CLOCK FAMILY)

NAME *Boerhavia coccinea*

A decumbent perennial with branching stems. *Leaves* Opposite, broadly triangular, pale green or grey beneath, with entire margins. *Flowers* Clusters of pink flowers on a long stalk, both terminal and axillary.

Flowering	Early dry season.
Habitat	Dry, sandy soils; areas 2 and 9.
Distribution	Pan-tropical.
Note	Roots have numerous medicinal uses, mainly for digestive disorders.

NAME *Commicarpus plumbagineus*

Scrambling herb with long branched stems; up to several metres. *Leaves* Slightly fleshy, more or less pubescent on both surfaces. *Flowers* White trumpet-shaped with long stamens. *Fruit* Wart-like sticky glands along the sides.

Flowering	Most of the year, if near water.
Habitat	Forest edges, near water; area 1.
Distribution	Tropical Africa, Palestine, S Spain and S Arabia.

Nymphaeaceae (WATER LILY FAMILY)

NAME *Nymphaea lotus*

Perennial aquatic herb; up to 45 cm. *Leaves* Float on water, round with serrated edges. *Flowers* White with yellow stamens, rise above the water, bloom at night and close mid-morning.

Flowering	Dry season and early rains, up to February.
Habitat	Slow moving, warm, slightly acidic water; area 8.
Distribution	Africa and SE Asia.
Note	Found in the tomb of Ramasses II in Egypt.

NAME *Nymphaea nouchali*

Perennial, clump-forming aquatic herb. *Leaves* Float on water, round, margins may be notched or smooth. *Flowers* White, blue or pink with yellow stamens.

Flowering	Dry season and early rains, up to February.
Habitat	Slow moving, warm water; area 8.
Distribution	Old World tropics.
Note	The national flower of Bangladesh.

Ochnaceae

NAME *Ochna pulchra*

Shrub or small tree with distinctive, grey, peeling bark; up to 4 m. *Leaves* Alternate, elliptic with entire margins, turn coppery brown in autumn. *Flowers* Yellow/green in many-flowered clusters, very short lived. *Fruit* Kidney-shaped, green turning black surrounded by conspicuous red sepals.

Flowering Late dry season to early rains.

Habitat Open woodland and sandy soils; area 9.

Distribution Central and southern Africa.

Note Seeds high in oil which are eaten or used to make candles and soap.

Olacaceae (AFRICAN WALNUT FAMILY)

NAME *Ximenia spp*

Ximenia americana

Ximenia caffra

A small deciduous tree with spine-tipped branchlets; up to 6 m. *Leaves* In dense clusters along the branches, leathery with entire margins. *Flowers* Small creamy green. *Fruit* Oval, bright orange to deep red with a single seed.

Flowering Late dry season.

Habitat Grassland & rocky outcrops; area 7.

Distribution Central and southern Africa.

Note Fruit is edible, having a refreshing sour taste. Ripe fruit has a vitamin C content of 27%, is high in potassium and contains protein. The seed has 65% oil content and is used to soften human skin and animal hides. It is also used for lamps.

Oleaceae (OLIVE FAMILY)

NAME *Jasminum fluminense*

Climbing shrub; up to 5 m. *Leaves* Opposite, compound with three triangular leaflets with entire margins. *Flowers* White with 5-8 free petals and a sweet scent. *Fruit* Black fleshy berries.

Flowering	Most of the year, if near water.
Habitat	River banks; areas 1, 8 and 9.
Distribution	Pan-tropical.
Note	*Jasminum* from Persian meaning Gift from God.

Onagraceae (EVENING PRIMROSE FAMILY)

NAME *Ludwigia leptocarpa*

Robust, hairy, shrubby perennial; up to 3 m. *Leaves* Alternate, crowded near top of stems, narrow with long hairs and entire margins. *Flowers* Solitary with five free petals. *Fruit* Dull brown, hairy capsule.

Flowering	Most of the year, if near water.
Habitat	Damp, open ground; area 9.
Distribution	Widespread in tropical Africa, a weed in damp places in tropical America.

NAME *Eulophia guineensis*

Terrestrial with clusters of pseudobulbs above ground; up to 45 cm. *Leaves* Basal, elliptic with entire margins, on stalks. *Flowers* Greenish/purple, lip large with wavy margins, white/pale pink with a darker purple blotch in the throat.

Flowering Early rains.

Habitat Shady places near rivers; area 1.

Distribution Africa and the Arabian peninsula.

NAME *Eulophia leachii*

Terrestrial herb, pseudobulbs mostly above ground; up to 1 m. *Leaves* Basal, linear with fine serrations. *Flowers* Sepals green, petals and lip maroon with green veins and a white tip, on a single spike.

Flowering Early rains.

Habitat Deciduous woodlands, sandy soil near water sources; area 8.

Distribution Lower altitudes in Zambia and Zimbabwe.

Oxalidaceae (WOOD SORREL, OCA AND STARFRUIT FAMILY)

NAME *Biophytum spp.*

Biophytum petersianum

Biophytum abyssinicum

An annual herb; up to 40 cm. *Leaves* A rosette of compound leaves, leaflets elliptic with smooth margins, final leaflet largest, sensitive to touch. *Flowers* Orange in the centre of the rosette *(B. petersianum)* or white on a long stalk *(B. abyssinicum)*. *Fruit* Capsule with long flattened seeds.

Flowering	Rains.
Habitat	Sandy soils and woodland; area 2.
Distribution	A common weed in the Old World tropics.
Note	Sap used to cure cutaneous and sub-cutaneous parasitic infections. In Mozambique a concoction is used for snake bite.

NAME *Oxalis corniculata*

Small, creeping through grass, rooting at the nodes. *Leaves* Opposite, compound with three leaflets, clover-shaped, on short stalks. *Flowers* Yellow, five petals, borne in the axils. *Fruit* Narrow cylindrical capsule.

Flowering	Rains.
Habitat	Disturbed ground; area 9.
Distribution	Cosmopolitan weed.
Note	Leaves edible with a tangy taste, rich in vitamin C. Contains oxalic acid which can inhibit the absorption of calcium. Oxalis means sour.

Papaveraceae (POPPY FAMILY)

NAME *Argemone mexicana*

Erect, spiny annual, stems with yellow latex; up to 1m. *Leaves* Alternate, blue/green with a distinctive white stripe down the centre and along the veins, deeply lobed with sharp, spiny margins. *Flowers* Bright yellow, papery. *Fruit* Oblong, erect, ribbed capsule with sharp prickles. *Seeds* black.

Flowering	Most of the year.
Habitat	Disturbed, open ground; area 9.
Distribution	Native to Mexico, now an invasive pan-tropical weed.
Note	Flower toxic, has narcotic effects.

Passifloraceae (PASSION FLOWER FAMILY)

NAME *Adenia digitata*

Herbaceous climber with stems up to 3 m; *Leaves* Alternate, deeply lobed with entire margins. *Flowers* Cream/yellow with red markings on the outside, a long corolla tube opening into 5 free petals, ending in a tendril.

Flowering	Early rains.
Habitat	Open woodland near rivers; area 1.
Distribution	Central and southern Africa.
Note	The tubers are very poisonous.

Pedaliaceae (SESAME FAMILY)

NAME *Ceratotheca sesamoides*

Slender erect annual, hairy; up to 60 cm. *Leaves* Opposite, arrow-shaped with serrated margins. *Flowers* Pink or mauve, throat cream with dark markings, pendent, borne in the leaf axils. *Fruit* A laterally flattened capsule with slender horns.

Flowering	Rains.
Habitat	Sandy soils in grassland or savannah woodland; area 9.
Distribution	Tropical and southern Africa.
Note	Leaves are edible. May be confused with *Sesamum angolense* which has narrower leaves.

NAME *Cerathotheca triloba*

Erect with a square stem, hairy; up to 1.3 m. *Leaves* Opposite, oval with broadly serrated margins. *Flowers* Pale pink with darker lines, resembling a foxglove. *Fruit* Pods with two horns.

Flowering	Early dry season.
Habitat	Sandy soils in open grassland or disturbed ground; area 9.
Distribution	Southern Africa.

Pedaliaceae (SESAME FAMILY)

NAME *Harpagophytum zeyheri*

Prostrate herb with annual stems.
Leaves Opposite, broadly oval with wavy margins. *Flowers* Pink with yellow throat, borne in leaf axils. *Fruit* Capsule with four lines of backward curving spines.

Flowering	Early rains.
Habitat	Dry open woodland, Kalahari sand; areas 4 and 5.
Distribution	Widespread in southern Africa.

NAME *Sesamum alatum*

A tall erect herb; up to 1.5 m.
Leaves Opposite, lower leaves 3-5 foliate, linear with entire margins, upper leaves simple. *Flowers* Dark pink with markings on the corolla lip, borne in the leaf axils. *Fruit* Capsule with seeds winged at both ends.

Flowering	Late dry season, early rains.
Habitat	Waste land; area 5.
Distribution	Tropical Africa and the Indian sub-continent.
Note	*alatum* meaning winged.

Phyllanthaceae

NAME *Phyllanthus reticulatus*

A much-branched, deciduous shrub; up to 3 m. *Leaves* Alternate, oval with entire margins. *Flowers* Yellowish, small, in clusters. *Fruit* Red, turning black, berry-like.

Flowering	Dry season.
Habitat	Hot, dry deciduous woodland; areas 1 and 8.
Distribution	Old World tropics and sub-tropics.
Note	Nicknamed the 'potato-bush', due to its characteristic smell of boiled potatoes.

Plumbaginaceae (LEADWORT AND SEA LAVENDER FAMILY)

NAME *Plumbago zeylanica*

A scrambling perennial, stems smooth at the base, with glands at top; up to 2 m. *Leaves* Alternate, trifoliate with entire margins. *Flowers* Funnel-shaped with five white petals above a sticky calyx. *Fruit* Small, cylindrical with pointed calyx.

Flowering	Most of the year, if near water.
Habitat	Damp, shady places; areas 1 and 8.
Distribution	Native to the Old World tropics, now pan-tropical.
Note	Has anti-microbial and anti-fungal properties.

NAME *Aristida congesta*

NAME *Chloris virgata*

NAME *Dactyloctenium giganteum*

NAME *Eragrostis superba*

NAME *Hyparrhenia hirta*

NAME *Melinis repens*

NAME *Miscanthus junceus*

NAME *Panicum maximum*

NAME *Phragmites mauritianus*

NAME *Sporobolus panicoides*

NAME *Tragus berteronianus*

Polygonaceae (BUCKWEED, RHUBARB AND DOCK FAMILY)

NAME *Oxygonum sinuatum*

Almost hairless, usually decumbent annual; up to 20 cm. *Leaves* Alternate, small. *Flowers* Single, white with five petals, borne singly along the stems. *Fruit* Triangular, with prickles.

Flowering Rains.

Habitat Common weed on waste ground and exposed rocky places; area 2.

Distribution Tropical Africa.

Note Leaves are edible, and can be boiled and pounded to form a poultice for skin lesions.

NAME *Persicaria senegalensis*

A robust perennial; up to 3 m. *Leaves* Alternate, bright green with entire margins. *Flowers* Pink clusters on long spikes. *Fruit* Shiny black nut.

Flowering Most of the year, if near water.

Habitat In standing water; areas 8 and 9.

Distribution Throughout Africa and west Asia.

Note Leaves are edible.

Pontederiaceae (WATER HYACINTH FAMILY)

NAME *Eichhornia crassipes*

Aquatic, floating perennial herb; up to 40 cm. *Leaves* Rosette with swollen stems and smooth margins. *Flowers* Mauve, the top petal has a yellow spot with purple surround, up to 20 flowers on an erect spike, close at night.

Flowering All year, in water.

Habitat Slow flowing water; areas 1 and 9.

Distribution Pan-tropical.

Note An invasive weed, very fast growing. Thrives in polluted water forming a thick blanket that removes oxygen and deprives water of light so killing other aquatic life. Chokes water ways and hydroelectric dams. Originally an escaped ornamental.

NAME *Clematis brachiata*

A deciduous climber reaching up to 5 m. *Leaves* Compound with 5-7 oval leaflets. *Flowers* Small, sweetly scented, creamy white, petals reflexed. *Fruit* Large fluffy seed heads which last several months.

Flowering Early dry season.
Habitat Damp areas alongside streams; area 9.
Distribution Tropical and southern Africa and Madagascar.

NAME *Ranunculus multifidus*

Perennial herb; up to 1 m. *Leaves* Mainly basal, trifoliate, hairy. *Flowers* Yellow on long branched stems.

Flowering Rains.
Habitat Damp grasslands; area 2.
Distribution Common in Africa, Arabia and Madagascar.

NAME *Carphalea pubescens*

Erect or scrambling shrub; height 60-350 cm. *Leaves* Opposite, lance-shaped with entire margins. *Flowers* White with a long tubular corolla and four to five spreading lobes. Surrounded by a conspicuous, asymmetrical, pale green calyx with green stripes. *Fruit* Ribbed with enlarged calyx lobes on top.

Flowering Rains.

Habitat Open woodland, Kalahari sand and rocky basalt; areas 4, 7 and 9.

Distribution Central Africa.

NAME *Catunaregam taylorii*

Shrub with spreading or trailing branches, spines mostly paired; up to 5 m. *Leaves* Opposite, broadly round with entire margins, hairy, immature at time of flowering. *Flowers* White maturing to yellow, fragrant.

Flowering Early rains.

Habitat Acacia woodland; area 6a.

Distribution Central and southern Africa.

Rubiaceae (COFFEE AND MADDER FAMILY)

NAME *Crossopteryx febrifuga*

Shrub or small deciduous tree; up to 6 m. *Leaves* Opposite, obovate with conspicuous net-veining below, margins entire. *Flowers* Cream in dense terminal heads, fragrant.

Flowering Early rains.

Habitat Sandy soil in dry mixed woodland; area 6.

Distribution Tropical Africa.

Note Bark used for treatment of diarrhoea and dysentery.

NAME *Feretia aeruginescens*

Scrambling shrub or small tree; up to 3m. *Leaves* Opposite, obovate with entire margins, appearing after the flowers. *Flowers* White/pink solitary or in clusters borne on bare branches. *Fruit* Spherical deep red when mature.

Flowering Early rains.

Habitat Woodland near water; areas 1, 8 and 9.

Distribution Central and southern Africa.

Rubiaceae (COFFEE AND MADDER FAMILY)

NAME *Gardenia resiniflua*

Shrub or small tree with grey, slightly peeling bark; up to 3 m. *Leaves* Whorls of three on short side branchlets, obovate, rough with entire margins. *Flowers* White, tubular, solitary in upper leaf axils. *Fruit* Almost round, yellow/brown densely covered in hairs.

Flowering Early rains.

Habitat Woodland and scrub – often along rivers; area 1.

Distribution Central and southern Africa.

NAME *Gardenia volkensii*

Densely branched shrub or small tree; up to 3 m. *Leaves* Whorls of three leaves on short side branches, rough, broadly ovate, with entire margins. *Flowers* White, showy, on ends of branches, fade to yellow. *Fruit* Grey/brown, almost spherical with warty lenticels.

Flowering Late dry season to early rains.

Habitat Open deciduous woodland, sandy soils; areas 1 and 8.

Distribution Tropical and southern Africa.

NAME *Pavetta cataractarum*

A shrub or small tree; up to 2.5 m. *Leaves* Opposite, with entire margins, velvety grey below. *Flowers* Large balls of tubular white flowers. *Fruit* Small round and velvety.

Flowering Rains.

Habitat Rainforest; area 1.

Distribution River banks in Zambia, Zimbabwe and Tanzania.

NAME *Tricalysia junodii*

Scrambling shrub; up to 2 m. *Leaves* Opposite, bright green with wavy margins. *Flowers* White often tinged pink, bell-shaped with white fluffy hairs in throat, in clusters of up to five hanging from long stalks. *Fruit* Almost round, hairy black when ripe.

Flowering Early rains.

Habitat Open deciduous woodland; areas 1 and 9.

Distribution Central and southern Africa.

NAME *Aptosimum decumbens*

Decumbent branching under-shrub spreading up to 1 m. *Leaves* Alternate, simple, oblong with entire margins. *Flowers* Bright purple, tubular, with small green bracts, borne in the leaf axils. *Fruit* A capsule.

Flowering Rains.

Habitat Open woodland, on soils ranging from sand to basalt; area 7.

Distribution Native to Central and southern Africa.

NAME *Buchnera hispida*

Hemi-parasitic erect annual, black when dry; up to 80 cm. *Leaves* Linear with entire margins. *Flowers* Purple at the end of branches in loose groups. *Fruit* Ovoid capsule.

Flowering Rains.

Habitat In waste and sandy places; area 7.

Distribution Throughout Africa, Madagascar and west Asia to India.

Note A hemi-parasite on grasses and grass-like crops.

Scrophulariaceae (FIGWORT AND FOXGLOVE FAMILY)

NAME *Craterostigma plantagineum*

Perennial herb; up to 10 cm. *Leaves* Basal rosette, broadly ovate to lance-shaped with entire margins. *Flowers* Purple and white with two lips, the lower lip three-lobed, centre lobe very broad.

Flowering	Early rains.
Habitat	Shallow soil overlying rock; area 2.
Distribution	Tropical and southern Africa.
Note	Can lie dormant for long periods until it rains.

NAME *Cycnium tubulosum*

Erect or straggling perennial herb; up to 80 cm. *Leaves* Opposite, linear, widely spaced on stems. *Flowers* White, pink or mauve, fragrant.

Flowering	Mainly in rains, may flower all year if near water.
Habitat	Damp grassland and river margins, may grow in water; areas 1 and 8.
Distribution	Common throughout Africa and Madagascar.
Note	Partial root parasite on grasses.

NAME *Datura inoxia*

Erect annual, covered in short white hairs; up to 1.5 m. *Leaves* Alternate, elliptic with wavy margins. *Flowers* White trumpets with mauve centres which open at night. *Fruit* Spiny egg-shaped capsule.

Flowering	Rains.
Habitat	Damp ground; area 1.
Distribution	Native to America, now pan-tropical.
Note	Poisonous, causes hallucinations.

NAME *Solanum nigrum*

An erect annual with slender branched stems; up to 60 cm. *Leaves* Alternate, oval with entire or slightly wavy margins, sometimes hairy. *Flowers* Five triangular petals, often curved backwards from a yellow centre. *Fruit* A berry turning from green to red and finally black as it ripens.

Flowering	Most of the year.
Habitat	Near water and around human habitation; area 9.
Distribution	Native to Europe, now widely naturalised.
Note	Crushed black berries are used as a skin dye. Toxic to livestock.

Solanaceae (POTATO FAMILY)

NAME *Solanum panduriforme*

A perennial herb or small shrub; up to 1 m.
Leaves Alternate, elliptic with wavy margins, pale underneath with grey hairs.
Flowers Purple with yellow stamens.
Fruit Round, mottled green, turning yellow when ripe.

Flowering Late rains.
Habitat Waste ground; area 9.
Distribution Tropical and southern Africa.

NAME *Solanum seaforthianum*

Vine; climbing to 5 m. *Leaves* Alternate, pinnately divided with unequal base. *Flowers* Bright purple with yellow stamens hanging in clusters. *Fruit* Clusters of green berries turning red as they ripen.

Habitat Common in woodland climbing over shrubs and anthills; areas 1 and 9.
Distribution Native to the West Indies, now widely naturalised in the tropics.
Note May be invasive.

Sparrmanniaceae (JUTE, SPARRMANNIA AND GREWIA FAMILY)

NAME *Corchorus asplenifolius*

Very variable, perennial herb, stems may be hairy; up to 50 cm. *Leaves* Alternate, narrow with widely serrated margins. *Flowers* Yellow with five free petals and numerous stamens, borne in the leaf axils. *Fruit* Cylindrical capsule with many dark brown seeds.

Flowering	Rains.
Habitat	Open woodland and margins of seasonal swamps; area 9.
Distribution	Central Africa.
Note	Leaves are a popular food, ash can be used as a substitute for salt, also used to dispel ants.

NAME *Corchorus olitorius*

An erect woody herb; up to 1.5 m. *Leaves* Alternate, oval with finely serrated margins. *Flowers* Yellow, solitary, borne in the leaf axils. *Fruit* Cylindrical capsule splitting into five parts. *Seeds* grey/black, angled.

Flowering	Rains.
Habitat	Open grassland; area 5.
Distribution	Native to India/Pakistan, now widespread in the tropics.
Note	Leaves, which are high in protein, are used in salads and as a vegetable. Fibre from stems used for sackcloth.

Sparrmanniaceae (JUTE, SPARRMANNIA AND GREWIA FAMILY)

NAME *Grewia olukondae*

A shrub, larger branches are often square; up to 2 m. *Leaves* Alternate, oval with irregularly serrated margins, hairy on both sides. *Flowers* Yellow, star-shaped with reflexed petals. *Fruit* Orange/brown, hard, berry-like with two seeds.

Flowering	Rains.
Habitat	Along river banks, often in dense stands; **areas 1 and 8.**
Distribution	East, Central and S Africa, and India.
Note	Fruit eaten by humans, hornbills and other birds, leaves browsed by impala.

NAME *Grewia pachycalyx*

Many-stemmed scrambling shrub or small tree; up to 5 m. *Leaves* Alternate, oblong, often drooping, with fine serrations. *Flowers* White, showy in leaf-opposed clusters. *Fruit* Four lobed, orange/red when ripe.

Flowering	Early rains.
Habitat	Open grassland, near rivers; **areas 1 and 8.**
Distribution	Central Africa.

Sterculiaceae (COLA AND STERCULIA FAMILY)

NAME *Melhania forbesii*

Branching shrublet covered in short hairs; up to 60 cm. *Leaves* Alternate, oval, margins have broad serrations. *Flowers* Up to three at the leaf nodes, with hairy buds. *Fruit* Dark brown.

Flowering Rains.

Habitat Open grassland; areas 7 and 9.

Distribution Old World tropics.

Note Each flower opens for only one day – usually only in the morning.

Turneraceae

NAME *Tricliceras glanduliferum*

An erect annual with hairy stems; up to 40 cm. *Leaves* Opposite, hairy with serrated margins. *Flowers* Yellow, single, terminal with five petals. *Fruit* A hairy capsule.

Flowering Rains.

Habitat Rocky and sandy savannah; area 5.

Distribution Common in southern and Central Africa.

Typhaceae (CATTAIL FAMILY)

NAME *Typha capensis*

Robust aquatic perennial; up to 2.5 m. *Leaves* Long strap-shaped with entire margins. *Flowers* On terminal spikes, male short-lived on slender spike above, female below thicker with densely packed flowers, turning velvety brown as they age. *Fruit* Minute, one-seeded with many hairs.

Flowering	Early rains.
Habitat	Dams and water courses, often forming large stands; area 9.
Distribution	Cosmopolitan.
Note	All parts are edible. Used medicinally in child birth and used to treat sexually transmitted diseases. Provides shelter for many birds and small aquatic animals.

Vahliaceae

NAME *Vahlia capensis*

Annual or perennial shrub with erect stems; up to 50 cm. *Leaves* Opposite, linear with entire margins. *Flowers* Yellow in axillary pairs with sharply pointed sepals.

Flowering	Late dry season to early rains.
Habitat	Open areas on sandy soil, often near water; area 5.
Distribution	Old World tropics, excluding the Arabian peninsula.

Velloziaceae (VELLOZIA FAMILY)

NAME *Xerophyta equisetoides*

Erect stems with a few short, stubby branches; up to 2 m. *Leaves* Clusters, emerging from the end of branches, linear with entire margins. *Flowers* Pale mauve or white, fragrant, on slender stalks.

Flowering	Early rains.
Habitat	Rocky hillsides; areas 3, 4, 5 and 6.
Distribution	Central and southern Africa.
Note	Dried leaves used as brooms.

NAME *Xerophyta humilis*

Low-growing perennial, sometimes mat-forming; up to 15 cm. *Leaves* In a basal tuft, linear with entire margins. *Flowers* On long stalks, pale mauve, with six free petals.

Flowering	Rains.
Habitat	On shallow soils in mopane-acacia woodland; areas 2 and 9.
Distribution	Southern Africa.
Note	From the Greek 'xeros' meaning dry, able to withstand long periods of drought. *humilis* meaning low-growing or dwarf.

Verbenaceae (VERVAIN FAMILY)

NAME *Lantana camara*

A small, aromatic shrub with square stems; up to 2 m. *Leaves* Opposite, simple, hairy, oval with serrated margins and an unpleasant smell when crushed. *Flowers* Dense heads of small, usually yellow and orange flowers which change shade after pollination. *Fruit* Green, turning black on maturity.

Flowering	All year.
Habitat	Very widespread, especially at the Falls; area 1.
Distribution	Native to the West Indies, now a pan-tropical weed.
Note	A rampant invasive plant, originally introduced as an ornamental. It should not be propagated. Poisonous to cattle.

Vitaceae (GRAPEFRUIT AND VIRGINIA CREEPER FAMILY)

NAME *Ampelocissus africanus*

Herbaceous liana, stems arising from a large semi-succulent perennial root. *Leaves* Alternate, large, palmate with small serrations. *Flowers* Small, dark red bunches, opposite the leaves. *Fruit* Round, green berries.

Flowering	Rains.
Habitat	Rainforest; area 1.
Distribution	Old World tropics.
Note	Fruit is edible.

Zingiberaceae (GINGER FAMILY)

NAME *Siphonochilus kirkii*

Perennial herb growing in groups; up to 30 cm. *Leaves* Basal, clasp the stem, ribbed, elliptic with entire margins. *Flowers* Pink, showy with white throat and yellow nectar guide. Up to 15 per plant, flowering in succession.

Flowering	Early to mid-rains.
Habitat	Semi-shade; area 9.
Distribution	Sudan, East and Central Africa.

Zygophyllaceae (CREOSOTE BUSH OR LIGNUM-VITAE FAMILY)

NAME *Tribulus terrestris*

Woody, prostrate, hairy perennial with branching stems; up to 60cm. *Leaves* Opposite, compound with up to seven pairs of oblong leaflets. *Flowers* Solitary, pale yellow with five free petals, on stalks arising from the nodes. *Fruit* Hairy, consisting of five segments, each with four distinctive spines.

Flowering	Throughout the year.
Habitat	Sandy soil, usually near habitation; area 9.
Distribution	Pan-tropical weed..
Note	Used as a treatment of oral infections in N Africa.

NAME *Actiniopteris radiata*

Small fern with radiating fronds; up to 15 cm.

Habitat Rocky outcrops; areas 2 and 3.

Distribution Tropical Africa, Arabia, India and Sri Lanka.

NAME *Nephrolepis cordifolia*

Bright green fronds; up to 80 cm.

Habitat Wet evergreen forest; area 1.

Distribution Tropical Africa and tropical America.

NAME *Selaginella imbricata*

Branched stems arise from a creeping stolon; up to 30 cm. *Leaves* Bright green, oval.

Habitat Damp rocky ledges and cracks; areas 2 and 3.

Distribution Throughout the tropics.

Note The dried plant re-grows as soon as the humidity reaches a sufficient level.

References

Biegel H. M. and **Tredgold M. H.** (1979). Rhodesian Wild Flowers. Thomas Meikle Series Number Four, The National Museums and Monuments of Rhodesia, Salisbury.

Blundell M (1992). Collins Photo Guide to the Wild Flowers of East Africa. Harper Collins, London.

Bolnick D. (1995). A Guide to the Common Wild Flowers of Zambia and Neighbouring Regions. Macmillan Publishers (Zambia) Ltd., Lusaka.

Bornman H. and **Hardy D.** (1971). Aloes of the South African Veld. Voortrekkerpers, Johannesburg.

Fanshawe D.B. (1972). The Vegetation of the Victoria Falls and its Environs. Forest Research Pamphlet no 45.

Heywood V. H., Brummitt R. K., Culham A. and **Seberg O.** (2007). Flowering Plant Families of the World. Royal Botanic Gardens, Kew.

Phillipson D. W. (ed) (1990). **Mosi-oa-Tunya**: A Handbook to the Victoria Falls Region. Longman, Zimbabwe.

Polhill R. and **Wiens D.** (1998). Mistletoes of Africa. RBG, Kew.

Reynolds G.W. (1996). Aloes of Tropical Africa and Madagascar. Trustees of the Aloe Book Fund, Swaziland.

Martineau R.A.S. and **Phear M. H.** (1953). Rhodesian Wild Flowers. Thomas Meikle Series Number One, The National Museums and Monuments of Southern Rhodesia. Longmans, Green and Co Ltd., Salisbury.

West O. (1974). A Field Guide to the Aloes of Rhodesia. Bundu Series, Longmans, Rhodesia.

Van Wyk B. E. & **Smith G.** (1996). Guide to the Aloes of South Africa. Briza Publications, Pretoria.

Main web sites used

www.aluka.org

www.apps.kew.org/efloras/fz

www.ePic.kew.org

www.fao.org

www.hear.org

www.ipni.org/index.html

www.plantzafrica.com

www.zimbabweflora.co.zw

Glossary

Alternate Leaves arranged alternately on the stem, not opposite.
Annual A plant that completes its life cycle in one year or one growing season.
Anther The terminal part of the male organs that carries pollen; usually on a stalk (filament).
Ascending Directed upwards at an oblique angle.
Axil/axillary The angle between the stem and leaf from which buds of new leaves or inflorescences grow.
Basal Originating at the base.
Beak A pointed projection.
Berry A fleshy fruit with the seed(s) immersed in pulp.
Bi-lobed With two lobes.
Blade The expanded part of the leaf usually attached to the stem by a short stalk.
Bract A modified leaf attached below a flower, often brightly coloured.
Calyx The outer whorl of floral parts, usually green, that protect the flower bud.
Capsule A dry, splitting fruit with two or more seed-containing compartments.
Compound leaf A leaf divided into a number of smaller leaflets (opposite of a 'simple leaf').
Corona A crown-like outgrowth from a corolla tube.
Corolla A collective term for petals which may be free or fused into a tube.
Cosmopolitan Worldwide distribution.
Deciduous Shedding leaves and/or other organs annually.
Decumbent Running along the ground, with the end ascending.
Dehiscent Splitting open.
Elliptic A leaf shape that is much longer than it is wide.
Ellipsoid Elliptical with a three dimensional body.
Endemic Restricted to a specific area – not occurring elsewhere.
Entire Without any breaks or serrations to leaf margins.
Epicalyx An extra whorl of appendages below the calyx.
Epiphyte A plant growing on another plant but not using the host plant for food.
Follicle A dry fruit with only one seed compartment splitting along the suture.
Gland A small superficial organ that secretes oils and other substances.
Herbaceous Not woody; mainly small annual or perennial plants.
Indehiscent Not splitting open at maturity (fruit).
Inflorescence A cluster of flowers.
Latex A fluid produced by certain plants, which contains various substances such as alkaloids, minerals, etc.
Leaflet A leaf-like segment of a compound leaf.
Lenticel A small elliptical pore.
Lip/labellum A single petal that is distinct from, and often larger than, the other petals, giving a flower a characteristic form; mainly in orchids.
Lobe(d) A division of an organ; generally rounded.
Mid-vein The main central vein of the leaf that is a continuation of the veins in the petiole.
Mucilaginous Slimy.
Native Belonging to a region through natural circumstances (not introduced).
Naturalised Introduced into a region and subsequently surviving naturally.
Node Point of attachment of leaves to the stem.
Nutlet A small nut.
Obovate Oval but with the narrow end attached to the stalk.
Opposite With leaves arranged as pairs on opposite sides of the stem, arising from the same node.

Orbicular Round.
Ovoid Egg-shaped.
Palmate (Leaves) With four or more leaflets arising from a single point. (Veins) Several equally prominent veins branching out from the base of a leaf blade.
Parallel venation Leaf veins lie parallel to each other.
Parasitic Deriving nutrients from another plant.
Pendent Hanging down.
Perennial A plant with an indefinite life span, taking more than a year to grow and which flowers for a long period of time.
Petiole The stalk of a leaf.
Pinnate With two rows of leaflets, one along either side of a mid-rib, feather-like.
Pod A dry splitting fruit, opening along two sutures.
Prickle A sharp pointed outgrowth from the superficial tissues of the stem.
Prostrate A plant that grows closely along the ground.
Recurved Curved backwards.
Reflexed Bent backwards.
Resin A sticky, often fragrant substance that is insoluble in water, secreted by specialised cells.
Rhizome An underground stem that grows horizontally.
Riverine Growing alongside a river.
Rosette A tight spiral of basal leaves.
Sap The juice of a plant.
Scale A reduced leaf.
Sepal A segment of the calyx.
Serrated With a saw-toothed margin.
Shrub A perennial woody plant with several main stems.
Simple leaf A single leaf arising from the axil (opposite of 'compound leaf').
Species A group of plants of constant distinctive characteristics. Linnaean unit of classification
Spike Flowers arranged on a central stem without stalks.
Spine A sharp woody outgrowth from a leaf.
Stamen The male part of a flower consisting of stalk (filament) and anther.
Staminode A sterile stamen.
Stellate Star-shaped.
Stem The main supporting axis of a plant.
Succulent A plant with fleshy leaves.
Tendril A thread-like structure that grasps other plants or objects for support.
Tepal Often strongly coloured segment of a flower, not differentiated into petals and sepals.
Terminal At the end of a stem or at the extreme top.
Thorn A short pointed branch.
Tree A perennial woody plant, at least 2m tall, with a main trunk.
Trifoliate With three leaves.
Tuber An underground stem or root that is often swollen with food reserves.
Tubular Cylindrical or hollow.
Venation Arrangement of veins in a leaf.
Whorled A circular arrangement of leaves with more than two to a node.
Xerophytic plant A plant which can live in very dry conditions.

Index

Index

Also by Helen Pickering:

Wild Flowers of Mombacho, Nicaragua.

Published 2006 by the Brit Press, 409 Pecan Street,
Forth Worth, Texas, 76102-4060, USA
ISBN 1-889878-14-6 ISSN 0833-1475
www.brit.org

By Helen Pickering and Annette Patzelt:

Field Guide to the Wild Plants of Oman

Published 2008 by Royal Botanic Gardens, Kew,
Richmond Surrey TW9 3AB, UK.
ISBN: 978 1 84246 177 8
www.kewbooks.com